Living a Sustainable Lifestyle for Our Children's Children

Living a Sustainable Lifestyle for Our Children's Children

R. Warren Flint and W.L. Houser

Authors Choice Press

San Jose New York Lincoln Shanghai

**Living a Sustainable Lifestyle
for Our Children's Children**

Authors Choice Press
an imprint of iUniverse.com, Inc.

For information address:
iUniverse.com, Inc.
5220 S 16th, Ste. 200
Lincoln, NE 68512
www.iuniverse.com

ISBN: 0-595-20013-3

Printed in the United States of America

Epigraph

"We are the whole reason why our ancestors have existed at all. Let our children's children be the reason why we exist, and ultimately act."

Words of Cinque from the book *Amistad*
by Alexs Pate, 1997; Dreamworks.

Contents

"Every animal leaves traces of what it was; man alone leaves traces of what he created."

Jacob Bronowski

List of Illustrations xi
List of Tables xiii
Preface: Sustainability xv
Acknowledgements xxiii
Introduction: Why this book? xxv

Section I: The Big Picture
Chapter 1 What is Sustainable Development? 3
    IMPROVING QUALITY OF LIFE *4*
    HITTING LIMITS *6*
    WHAT SUSTAINABILITY IS NOT *8*
    SUSTAINABILITY'S THREE BASIC ELEMENTS *10*
Chapter 2 Our Ecological Footprint **15**
    CONSUMPTION *16*
    HOW BIG IS YOUR FOOTPRINT? *19*
    WASTE *21*
    FORESTS *22*
Chapter 3 Beyond the Obvious **25**
    EXTINCTION REPEATED *26*
    ACCELERATING CYCLES *28*
    SUCH A FINE LINE *31*

Section II: The Devil is in the Details

Chapter 4 The Chemical Story **37**
  A HISTORY *38*
  PERSISTENCE IS EVERYTHING *39*
  THREATS TO FUTURE GENERATIONS *42*
  STABILITY AND MAGNIFICATION *46*
  UNANTICIPATED CONSEQUENCES *47*
Chapter 5 The Way of Nature **50**
  PATTERNS IN GENERAL *50*
  HOW IT ALL FITS TOGETHER *52*
  NATURE IS NOT LAWLESS *55*
  NATURAL CAPITAL *58*
  WASTE TO FOOD *61*
  NATURE'S APPROACH TO INSURANCE *62*
  UNDERSTANDING OURSELVES, OUR EARTH *65*

Section III: Why are People and the Earth at Risk?

Chapter 6 Outside the Web of Life **71**
  OUTGROWING EARTH'S CAPACITY *71*
  THERE IS NO SUBSTITUTION FOR NATURE *75*
  IS CLIMATE REALLY CHANGING? *77*
  AWARENESS AND SUSPICION *83*
  AN EVOLVING SEPARATION *85*
  THE GREATEST ILLUSION:
    MIND-HEART SEPARATION *90*
  THE ETHICS OF OUR FAILURE *92*
Chapter 7 Traditional Economics **94**
  THE INDUSTRIAL AGE *95*
  INDEPENDENCE AND VICIOUS CIRCLES *101*
  INABILITY OF MARKET SYSTEMS *102*
  THE PROBLEM OF ALL GOODS
    AND NO BADS IN ECONOMICS *106*
  CONSUMERISM: THE SECOND HALF
    OF THE POPULATION PROBLEM *109*
  AN ECOLOGICAL ECONOMICS *111*

CULTURAL REVIVAL TOWARDS LEAN,
    EFFICIENT ECONOMIES *113*
Chapter 8 Excuses for not Trying Sustainable Development....117
    BLURRING THE ISSUES: ARGUING OVER DETAILS 118
    THE ROAD TO HAPPINESS *120*
    MYTHS *124*
    DO YOU FEEL ALONE? *127*

Section IV: How Do We Act Sustainably?
Chapter 9 What is Quality of Life? **133**
    HUMAN DIMENSIONS *134*
    ETHICS and VALUES *136*
    CULTURAL CAPACITY *139*
    DE-MYSTIFYING SUSTAINABLE DEVELOPMENT *140*
    PLAIN LIVING: LOCALIZE RATHER
        THAN GLOBALIZATION *145*
Chapter 10 Science, People, and Community **149**
    UNDERSTANDING THE VOICE OF SCIENCE *150*
    REFRAMING THE ISSUE *152*
    PEOPLE CONNECTIONS *155*
    DEEPENING PEOPLE'S SENSE OF COMMUNITY *157*
    WHAT PAIR OF GLASSES ARE
        YOU LOOKING THROUGH? *160*
    TRANSFORMING COMMUNITY INTERACTIONS *162*

Section V: Sustainable Development in Action
Chapter 11 Where Do We Start? **171**
    TAKE A FIRST STEP *172*
    SUSTAINABILITY ANALYSIS *174*
    LOCAL MARKET ECONOMY *182*
    COMMUNITY, INSTEAD OF GLOBALIZATION *186*
    APPLYING A MEASURING STICK CAN HELP *187*
    THE MONHEGAN ISLAND STORY *192*
    COMMITMENT TO LIFESTYLE REDESIGN *194*

Chapter 12 Sustainability in Motion: Some Solutions **198**
    INDIVIDUAL ACTIONS: AWARENESS…
          UNDERSTANDING…MOTION *199*
    THE POWER OF ONE *221*
    USING COMMON SENSE
          AND DOING GREAT THINGS *223*
    SOCIETAL ACTIONS:
          THE NEXT INDUSTRIAL REVOLUTION *225*
    COPYING NATURE IN INDUSTRY:
          AN INTRIGUING IDEA *228*
    READY TO TAKE ACTION? *233*
Epilogue **237**
About the Authors **239**
Appendix I: Pertinent Internet Web Sites
          on Sustainable Development **241**
Bibliography **249**
Endnotes **255**

List of Illustrations

Figure 1.   Diagrammatic representation of the sustainability model showing the interconnection among community sectors that should occur in planning projects.
....11

Figure 2.   The directionality that guides cross sector thinking when considering the hierarchy of economy, society, and environment issues in evaluating the sustainability of any action or project.
....13

Figure 3.   Diverse plant assemblages and massive root system of a tree in a Jamaica rainforest.
....23

Figure 4.   The location of this cemetery adjacent to a chemical manufacturing plant along the Mississippi River in Louisiana is ironic with regards to the risks that residents face living in this region, often referred to as "Cancer Alley."
....43

Figure 5.   Steel production plant on the shores of Lake Ontario, Canada, emitting large amounts of smoke into the atmosphere that contain both carbon dioxide and sulfur dioxide, contributing to the green house gases above the Earth.
....57

Figure 6.   One day you pass this forested location and there is a wonderful wooded hillside habitat. A month later this is what is left after a logging company has clear cut the area for wood.
....66

Figure 7.   The millenium cooling trend suddenly reverses (from The Mercury's Rising, 1(1), 1999). In the last 1,000 years the average annual temperature of the globe's Northern Hemisphere has been dropping slowly, that is until 1900. Since then there has been a dramatic temperature rise, with 1990 the warmest year known since 1400 AD. The plot above shows the average movement (anomaly) of annual temperature away from a 1,000 year mean (represented by 0.0 on the X-axis).
....79

Figure 8.   The web of life in nature is as detailed, complex, and sensitive as a spider web you would find woven between two trees in the woods, as this one on Virginia's Eastern Shore.
               89

Figure 9.   Excessive logging, environmentally unsound agricultural practices, and other poorly designed building modifications to the landscape can create serious risks for erosion of land to occur.
               100

Figure 10.   When an oil spill occurs, such as the IXTOC I oil well blowout in the Gulf of Mexico in 1979 (above), the costs for environmental cleanup and remediation from this damage are counted as an economic good (on the income side) instead of an economic bad (on the cost side) as far as the Gross Domestic Product is concerned. The lighter color on the water's surface in the photo is the oil spewing from the well blow out located in the middle of the photo.
               108

Figure 11.   The interdisciplinary, cross-boundary approach to considering sustainable development requires examining all the various elements of a community's ecologic objectives, social objectives and economic objectives simultaneously.
               142

Figure 12.   A conceptual scheme that demonstrates the difference in the way society responds to the "symptoms" of a problem versus the way society should respond to the "causes" of that same problem instead.
               143

Figure 13.   Intensive corporate farming in action. A worker spraying massive amounts of pesticides on rows of tomatoes in plasticulture on the Eastern Shore of Virginia (USA).
               179

Figure 14.   Factories are an extremely large source of air pollution and gases that contribute to the greenhouse effect around the world. In addition, automobiles made by these factories are an equally important source of air pollution.
               201

Figure 15.   The Fractal Ecology Triangle conceptualized by Bill McDonough for explaining the balance that must exist between ecology, equity, and economy in any industrialized society for it to ultimately be sustainable.
               231

List of Tables

Table 1.   Differences between the indigenous Native world view and the western world view.
....161

Table 2.   Comparison of characteristics described by the early 20th century Industrial Revolution, the Industrial Revolution with the incorporation of the idea of "eco-efficiency", and the Next Industrial Revolution based upon the concept of eco-effectiveness.
....229

Preface

Sustainability

"The person who can begin early in life to think of things as connected, even if she or he revises their view with every succeeding
year, has begun a life of learning"
(Mark Van Doren, 1943)

Sustainable development—a way to live comfortably within the con-
fines of one's economic, environmental, and social limits—has become
a widely recognized goal for our society ever since deteriorating environmental conditions in many parts of the world began to suggest we
could be facing difficult survival problems. Modern society faces the
ruination that once brought down seemingly invincible civilizations in
the past. Then, the collapse was comparatively local in scale; today it
is global.

A shift to sustainable development is starting as a new reality is
being perceived by people; many of us are beginning to understand
that the reality we now hold is not our own. With the absurd use of
propaganda, commercialism, and subliminal influence the perception
of most of society, one shaped socially and politically outside a specific circle of control and manipulated by a few at the expense of the
many, is showing it's faults.

Though manipulative hierarchies are as old as mankind, we are
now faced with an era that can no longer environmentally or ethically
tolerate the influx of fallout from this fatalistic egocentrism. There is

nothing more blatant than environmental genocide. There is nothing more tragic than watching that which you love slowly deteriorate due to environmental poisoning. As with *Culture Jam, Last Hours of Ancient Sunlight, A Post-Corporate World,* and many others, this book carries the message that, **"We *can* change the world**" as individuals.

*   *   *

It is ordinary people living ordinary lives who are the heartbeat of the world society. It is the factory workers, the construction hard-hats, the paper pushers, and the bargain makers working day in and day out that keep the great wheels of progress rolling. It is the phone operators, the mall clerks, the hotel managers, and the truck drivers that keep the unending flow of commerce moving. It is the kindergarten teachers, the social workers, the mental health counselors, and the low profile environmental inspectors that put band-aides over the wounds left when one is run over, worn out, or almost destroyed by these wheels of progress. And, it is exactly those of us who are the pulse of the world, who are also becoming the heartbeat of sustainable development.

Because of this vital role, we hope to assist the advancement of an alternative view of the world that offers a better understanding for the many interconnections that exist, thereby improving our well-being and the well-being of our communities and environments. Lifting the mystery that surrounds the ramifications of consumer-driven lifestyles could be the saving grace of our planet and therefore our lives, as well as our children's lives.

Going beyond science, technology, and politics, this book is about day to day, moment to moment decision making. It brings awareness of how we live and why we live the way we do, which goes into the minds and hearts of all of us, addressing the basics of life: how to

know what is in our water, air, food, and land. It offers less dependence on that winning lottery ticket and more focus on working
towards a communal sense of safety, security, and relief. And we ask a
very basic, but strategic question: **what are we leaving our children?**

*   *   *

In *The Nature of Economies* Jacobs suggests that together plants and
animals compose an ecosystem, similar to a collection of businesses,
where their actions compose a community's economy.[1] There is no
predictable, hierarchical command over the ecosystem, but rather it is
making itself up as it goes along, being shaped by the many coincidences that may occur and the various responses to these coincidences
that ensue.

For example, all life, plants and animals alike, follow a road that
leads to the future. They do this without consciously thinking about
the future but rather just conducting their lives, including the response
to every event that happens to them, such as being chased by something wanting to eat them, carried on the wind to germinate in a new
place (plant seeds), or discovering an injured antelope that will make
a meal. Thus, development in nature is not about a collection of things
but rather about a process that yields things. And this process can be
influenced by a myriad of interactions, with many different resulting
outcomes.

Thus, nature allows for the development of many different
ecosystems with differing purposes that serve all of life in a very
successful integrated way. We know this works because nature's
integrity has supported all of life for millions of years. Like nature,
economic life permits us to fill material needs by developing cultures
and multitudes of purposes. But the ways these might be ultimately
achieved are by no means predictable. And in fact, are probably more

influenced by chance events than we would ever comfortably acknowledge. In the end, natural order as well as economic order, will occur through unpredictable self-organization,[2] evolving through the many different coincidences that happen.

*   *   *

Why are we drawn to the forests or the ocean? What is the draw to mountains or even deserts for some? Nature is quite a seductress. Quietly directing life she understands the duality of modernism as she undermines old concepts of materialism and transforms them into something more Earthly and meaningful. There are many interconnections and cycles that support and direct our life path via nature. At a genetic, almost cellular level we know this and follow. Our lives are often mysterious and unpredictable, but our beliefs, decisions, and actions do color the experiences that lie ahead.

Prince Charles of England probably said it best when he stated that sustainable development should be based upon a combined view of the "essential unity and order of the living and spiritual world," rather than on an omnipresent world view that seeks to reduce nature "to the level of nothing more than a mechanical process."[3] If there is not a spiritual awareness for the sacred coincidences that drive the natural world, what is to prevent humans from treating our entire world as a large laboratory of life with potentially unforeseen, disastrous consequences? The lack of a sense of connectedness allows the ego full rein, causing some to believe they run the world.

These beliefs lead to the inability to accept that there are limits on human ambitions. Because societal well-being is measured in terms of money, it is difficult to see the need to be careful with our environments. Though evidence is everywhere as to environmental decline, the degree of energy used to support habits and lifestyles—maintaining the illusion of wellness—is where the real message is missed. Is it possible to develop a precautionary posture?

*   *   *

Cautionary interaction with the natural environment is gaining public support but still faces significant official opposition. This block is born from the world view that nature's processes are simply a sum of those parts that have value or produce the most cash. This view blocks our ability to be guided by a spiritual awareness that we, nature, and all of its coincidences are connected.

Awareness is a balance between reason and intuition. Only by applying both the intuitive awareness from our hearts in an equal way with the reasoning of our minds, will we be able to appreciate the sacred guardianship for the world's environment offered us by a grander power. Practical and intuitive learning from our past can be blended with appropriate technology and knowledge of the present to produce the type of modern citizen that is acutely aware of both the visible and invisible worlds that inform our most successful actions in life.[4]

Acknowledging life's coincidences, responding to our consciousness, will bring us the **understanding** we need. Thus, the processes of discovery, while accepting and capitalizing on the sometimes mysterious coincidences that guide us, is dependent on our ability to remain open to new things and find deeper meaning in all events and interactions. The process and quality of our lives, our children's lives, and our grandchildren's lives depends on our ability to see beyond the haze of commercialism that has taken the world by storm.

Over 100 years ago, the great Indian Chief Seattle was faced with the loss of his tribe's land. He responded out of his love and respect for the land. He trusted the Chief in Washington, DC to do right by his land. He trusted in human nature and human awareness, believing that these would bring a real understanding to the tribe's values. And, he warned of the repercussions that walk hand in hand with loss of respect for nature. In a speech that encapsulated heart breaking eloquence he stated: "Whatever befalls the Earth…. befalls the sons and daughters of the Earth…. We did not weave the web of life; we are merely a strand in it. Whatever we do to the web, we do to ourselves." Therefore, we are directly responsible for the actions contained in our **motion**.

*   *   *

The decisions humankind makes over the next two decades are likely to determine whether or not the Earth life-support systems are sustained or become irreversibly impoverished. Without careful thought, both about deeper values and goals as well as appropriate policies and strategies, the best endeavors are likely to go round in ever decreasing circles.

We are only dimly aware of the future effects of our present actions, as we do not know for example, the full consequences of losing a species or particular kind of habitat like that marsh we really want to build over. That is why we, and all future generations, must take greater responsibility for our actions. Our challenge is to transform our newfound **awareness** of human-environment interactions into a deep commitment that further **understanding** can bring toward the development of positive **motion**, allowing us to both protect and wisely use our global assets.

The real challenge is that we be open to a spiritual awareness that can be gained by acknowledging life's coincidences, while constantly trying to understand the new and unpredictable instead of hiding from it under the guise that we think we already have the solution for any problem that may exist. The world, natural and man-made, is built to respond to our consciousness, but will only show us the level of understanding we are willing to commit to. Thus, the processes of discovery, while accepting and capitalizing on the sometimes mysterious coincidences that can truly guide us, is dependent on our ability to remain open to new things and find a real meaning in all events and interactions.

The importance of this challenge to making decisions at an individual level has influenced the flow of this book around the presentation of **awareness, understanding,** and **motion** issues as the most valid process to lead society to solutions that have longevity: for our children's children. In the pages that follow, we offer numerous alternatives for the person wanting to consider their individual ability to make a difference. We present you with questions. The right answers to these questions are what you personally believe after considering the evidence offered and the different ramifications of alternative decisions discussed. Your challenge is to think about and plan for a future that will make things less troubling, both for your own family and the environment.

Acknowledgements

In order to present a complete picture of the many aspects of sustainable development, we have thoroughly researched and relied upon the writings and opinions of those who have long advocated sustainability. It is in their footsteps we are simply following, and we would like to express a great appreciation for their vision, their clarity, and the gifts of their many innovative thoughts. Our research includes excellent books by such noted authors as Chris Maser, Doug Muschett, Ted Bernard, Mathis Wackernagel, the late Donella Meadows, Thom Hartmann, Herman Daly, Bob Costanza, Jon Robbins, David Korten, Jane Jacobs, and others, written in a more technical and scientific context. Here we have adopted their ideas in an easy to understand language for everyone to appreciate.

If you wish to refer to any of these other books for additional information, please see our bibliography listing at the end of this book. In addition, because of the quickly evolving quantity of information on the Internet, at this book's end we have also assembled a listing of Internet resources on the topic of sustainable development.

Our two year writing effort has also benefited from review of this manuscript by a number of different people that have read various versions and provided their very helpful comments and insight. These contributions to enhancing our message have come from Dr. Jack Vallentyne, Ms. Pat Phillips, Mr. Charles Chapman, and Mr. Michael Rice.

Introduction

Why this book?

"We've made houses of hatred. It's time we made a place where people's souls may be seen and made safe."

Jewel

In the past there was little awareness of our effects on nature: mainly because populations were much lower, and the technical, chemical, plastic coated revolution was not in full swing. For example, in 1943 Thomas Watson, the then chairman of IBM, said "I think there is a world market for maybe five computers." What would he have to say in the new century where many households have home computers and 50% are connected to the Internet?

For earlier generations, perceptions of the future and all that it held were somewhat misconstrued. We had no idea what the future held for us or the problems our generation would be facing, let alone the avalanche effect that is occurring for our children and grandchildren. On the scale of our own lifetimes, change happens very slowly. Thus, the burden of dealing with the outcome of yesterday's and today's decisions about regional and global problems fall mostly on our children and grandchildren. This is one reason why the magnitude of real effects from growing populations has not yet taken hold in many sectors of society.

But population growth, so passionately promoted by many in previous generations, is now putting our children and grandchildren at risk.[5] For centuries parents have labored so their children might have better lives and more opportunities. We may now be doing just the reverse, guaranteeing that our children will not have the resources, opportunities, and healthy environments previous generations have enjoyed.

*   *   *

Until now, the most dramatic changes that our forefathers experienced were droughts, floods, famine, or war. But because of technology and cultural evolution all of this is changing. Some may say changing for the better because of new innovations in technology. Diseases are being vanquished, child mortality is falling, incomes are rising, and people are crossing oceans in hours.[6]

One can wonder though, what cost are these changes exerting on our future? Social injustice, economic exploitation, and environmental pollution are not natural. They are consequences of thinking that has molded human development for more than 200 years. Is this cultural evolution a blessing or a curse? Poverty, hunger, resource depletion, and global warming are not the problems, they are symptoms of something deeper, something as old as time. Greed, over-consumption, fear, and ignorance are, as always the real problems.

Our health is at risk due to the very quality of life that the technology, chemical, food, and plastic revolution bring. Since science and technology have made this growth possible, most believe that science and technology can make possible all the things we continue to want in the future. The expansion of global trade by new technology makes it easier and faster to get what we need from longer distances, however, serves to totally block our understanding for the real effects that increased numbers of humans have on the carrying capacity of Earth's resources to support all of us.

Today 72 square miles of good, productive land will be lost to encroaching deserts, the results of human mismanagement and over-population. Today the human population will increase by 263,000. And today we will add 2,700 tons of chloro-fluorocarbons (CFCs) to the stratosphere and 15 million tons of carbon to the atmosphere. For every 100 pounds of product manufactured in the U.S., 3,200 pounds of waste will be produced.[7]

Today alone tons of persistent chemical contaminants will be added to our atmosphere, water, and land. Over 100 square miles of rain forests globally will be eliminated. Between 30 and 85 species of plants and animals will be lost from the Earth. No one knows whether the count is closer to 30 or 85. Tonight the Earth will be a little hotter, its waters more polluted, and millions of our global neighbors a bit worse off.[8]

Today low levels of insecticides, weed killers, and fertilizers are commonly found in our water. The average American has 180 foreign chemicals flowing through his or her veins. Our grandparents had one, lead, a natural occurrence. As a direct result, today 1,400 people will die of cancer. All of these things are transient and dangerous,[9] ultimately contaminating and eliminating plants and animals at the rate of dinosaur extinction.

*   *   *

Yesterday, today, and tomorrow; by year's end the numbers regarding nature's losses are staggering. The total loss of rain forest will equal an area the size of the State of Washington. Expanding deserts will equal an area the size of the state of West Virginia. Seven to ten billion tons of carbon will be added to the atmosphere. The global population will have risen by more than 90,000,000. As of 2000, as many as

20% of the life forms that were on the planet in the year 1900 are thought to now to be totally eliminated.[10]

According to David Pimentel,[11] the World Health Organization reports more than 3 billion people are currently considered malnourished. This represents the largest number and proportion of malnourished humans ever in history. Deaths from malnutrition and other diseases have significantly increased, especially during the past decade, and there is no indication that this trend will decrease or reverse. What can we expect as population numbers continue to climb?

Grain production—which supplies 80% to 90% of the world food—has been declining since 1983 and should alert us to the potential for future food security problems and increasing malnutrition. If Americans would simply reduce their meat consumption by 10% there would be enough grain to feed 6 million people for a year. Yet, today 45,000 people will starve to death, 38,000 of those that die will be children.

About 1.2 acres of cropland per person is required to provide a diverse diet similar to that desired by the average American and European.[12] At present, this amount of land is still available in the United States for its current population. In contrast, worldwide only about 0.27 acres of cropland per person remains for food production. Since land is a finite resource, available cropland per person will continue to decline, both worldwide and in the U.S., as the human population increases. How quickly is the lack of fertile land going to out distance the world's need for food?

*   *   *

How far is our reach, the impression of a footprint we create, in the taking of Earth's natural resources? The average American uses 25 acres to support his or her current lifestyle. This corresponds to the size of 25 football fields put together. Europeans require nearly as much. In comparison, the average Canadian lives on a footprint 25 percent less and the average Italian on 60 percent less.

With a potential growth in global population of 10 billion by the year 2050, the average available space for producing food per person on Earth will be reduced to 0.3 acres. Already, preliminary estimates suggest the use of food, forest products, and fossil fuel consumption alone exceeds global carrying capacity by an unsustainable 30 percent, being larger than what the world has to offer as society consumes more and more. A time has arrived that thinking seriously about how activities impact our backyard is essential. And that backyard may extend around the globe.

Rainfall, as well as water captured in rivers and lakes, is essential for all plants, including agricultural crops. An interconnected sign to our growing global crisis with regards to supplying food includes the fact that as agricultural production increases to feed more humans, pressure on water supplies also increases. Growing an acre of corn to feed cattle takes 535,000 gallons of water, a luxury our world can't afford when that acre could be used to plant crops to feed people in a healthier, less destructive manner.

Because communities, states, and countries must share water, competition for water resources increases. There are so many danger signs in present day cultures around the world that should alert us. For example, farmers in the Indian state of Gujarat are continually drilling more irrigation wells, some 1.5 meters deeper than in the past. How many international conflicts are arising now over trans-boundary water rights? As a deterrent to potential problems, Canada and the U.S. jointly agreed in 1999 to ban any further export of water out of the international Great Lakes basin.[13]

Human activities are also reducing biodiversity throughout the world. Pollination of food crops, essential for one-third of the world's food supply and dependent on diverse species of pollinators, such as bees and butterflies, has been declining. Some U.S. crops already face serious problems due to lack of sufficient pollination. And to make matters worse, the genetically engineered crops that many farmers are growing appear to have deleterious effects on many of these natural pollinators.

*   *   *

Whatever we find ourselves facing today, be it disease, child abuse, crime, injustice, economics, energy shortages, lack of good jobs, extinction of species, poverty, destruction of forests, pollution, breakdown of families, armed conflict, or nuclear power, the dynamics are the same.[14] There continues to be a frustrating perception of crisis and a feeling of powerlessness on a very personal level. Some of this frustration comes from the large knowledge gaps between people affected by many of these issues and the science that often provides understanding and solutions.

For example, while environmental awareness is globally at a record high, the Earth's natural resources are being degraded and/or depleted at astounding rates. Lack of sufficient knowledge certainly contributes to this crisis. But there are other factors also causing people to not carry through with actions to protect and conserve our planet's natural resources. In fact, according to Elizabeth Booth sustainable use of resources will depend on three inter-related elements: (1) improved technologies that foster economic development alternatives, as well as protect the environment; (2) enforced policies and laws that regulate and support these technologies; and (3) changes in the actions and behaviors of individuals, groups, and organizations.[15] Understanding and working with people's behavior is the **starting point** in defining new ways for society to make sustainable use of resources. Once these behaviors have been defined, one can identify the factors that influence their development and formulate strategies in technology and policies or laws that are more effective because they address and build upon those specific behavioral factors.

Environmental problems definitely do challenge our imaginations, our self-understanding of how we fit into the global design, our values, our sense of moral obligation to future generations, and thus our behaviors, in more profound and subtle ways than most people realize. What sort of world are we headed toward? Human numbers are growing and our economy has reached a scale where it can cause the extinction of whole animal populations, clear-cut forests, pollute most water, dirty the global atmosphere, and produce chemicals that contaminate our blood. People have become too large a burden on the Earth and that fact threatens our survival.

There is also growing agreement that the decisions we make as a society at this critical point will determine the course of the future for quite some time to come. Looking at how we live, how that in turn affects nature, and how fundamental nature is to our existence, is a beginning. Many are realizing that without a sound environment to live in, there is little more than this chaotic snowball effect where we constantly dodge a monumentous object rolling out of control straight at us. And in most instances we find ourselves quite inept at stopping it.

Through this book, we hope to ask and assist you in answering: What is real? What is possible? What will guarantee a future for our children's children with the same opportunities and promises of today? Answers to these questions require us as citizens to act as detectives and explore what the true meaning is of a sustainable lifestyle. The process of sustainability gives people a say in the air we breathe, the water we drink, and the food we eat. Thus, aside from sustainability ensuring our children's future, it makes for a better present for all of us.

Section I

The Big Picture

THE PARADOX OF OUR TIMES

These are the times of tall men and short character, steep profits and shallow relationships. These are the times of world peace and domestic warfare; more leisure but less fun; more kinds of food, but less nutrition. These are the days of two incomes, but more divorce; of fancier houses, but broken homes. It is a time when there is much in the show window and nothing in the stockroom; a time when technology can bring these words to you, and a time when you can choose either to make a difference or toss them aside.

We have taller buildings, but shorter tempers; wider freeways but narrower view points. We spend more, but have less; buy more, but enjoy it less. We have bigger houses and smaller families, more conveniences, but less time. We have more degrees, but less sense; more knowledge, but less judgement; more experts, but fewer solutions; more medicine, but less wellness.

We have multiplied our possessions, but reduced our values. We talk too much, love too seldom, and hate too often. We've learned how to make a living, but not a life. We've added years to life, not life to years. We've conquered outer space, but not inner space. We've been all the way to the moon and back yet have trouble crossing the street to meet a new neighbor. We've cleaned up the air, but polluted the soul. We've split the atom but not our prejudice. We have higher incomes, but lower morals. We've become long on quantity, but short on quality.

Chapter 1

What is Sustainable Development?

"You wake up one day and realize your standard of living somehow got
stuck on survive."

Jewel

Sustainable development is the simultaneous consideration of environment, life, and human well-being. Sustainability implies long-lasting, continuing far into the future.

At the beginning of the 20th century, when all but the last chunks of wild country were gone in the U.S., a few scrambled to protect what was left, including Theodore Roosevelt. Conservation became the centerpiece of his new progressive movement, which aimed at redressing the social, economic, and political imbalances—the imbalances caused by rapid industrialization, urbanization, and the concentration of economic power within the formerly unrestricted corporations. Preservationists clamored to save what was left of the nations' wildlands. By the time they got to work in the last quarter of the 19th century, less than 20 percent of the continent's pristine land remained intact.[16]

In the 1970s it was becoming apparent that global populations, poverty, environmental degradation, and resource shortages were increasing at a rate that could not long be safely continued. Because of this awakening a landmark book, *Limits to Growth,* detailing the many limits to human growth on a global scale was published.[17] Initially this book caused discussions around the world, but ultimately most concluded that the idea of limits on humans was too terrible to be true. While the message of limits diminished, the reality of limits has become an increasing truth which people have had to deal with.

As a means of deflecting the continually reappearing message of limits, the word sustainable began to appear as an adjective to concepts of growth.[18] It derived from the idea of "sustainable yield" used to describe agriculture and forestry when harvesting is conducted in a way that it could be continued indefinitely. Because of the continuing mystery on how to really achieve a sustainable yield, and the continuing doubt it conveys, the best approach to discussing sustainable development has been to talk about the things we all know about and are comfortable with: such things as our homes, our children, our jobs, nature, the air we breathe, and the food we eat. These topics are what sustainability is about and they are all interconnected.

IMPROVING QUALITY OF LIFE

As the World Commission on Environment and Development set forth in 1987,[19] sustainable development is improving our life-enabling habits to meet our needs in the present without compromising the ability of future generations to meet their needs. Natural resources such as water, air, soil, plants, and animals are the basic assets upon which all life, human and otherwise, depends. Therefore, according to this definition it is unwise to use up these supplies or we will be threatening the security of others in the present and future.

Through design of sustainable lifestyles, everyone can enjoy the good life without jeopardizing the future of our children's children. If we think of development as assuring a dignified level of existence—serving our basic needs, such as food, clean water, shelter, clean air, clothes, friendship, diversity of tastes, beliefs, preferences, and personal growth—we would take a great step toward constructing a better and more sustainable world. Living in material comfort, peacefully with each other, within the means of nature is the common ground that most defines sustainability.

But, the term "sustainable development" itself can be ambiguous. Many identify with the "sustainable" part and hear a call for ecological and social transformation, a world of healthy environments and social justice. Others identify with "development" and interpret it to mean more sensitive economic growth, a significantly reformed version of the status quo. The minute one equates growth with the word development, however, a deeper insight allows us to begin seeing immediate problems. Sustainable growth implies increasing endlessly, which can also mean the growing quantity will tend to become infinite in size.

We all understand how this is not possible, especially when the term "sustainable growth" is composed of words that are an oxymoron. Distinguishing development versus growth also includes discussing sustainable **development**, not sustainable **growth**. Economist Herman Daly says that "sustainable development is development without growth—as growth means getting bigger while development means getting better."[20]

In theory this should not be too difficult. Each of us does much the same thing in the course of our individual lives. We grow early in life and when we reach adult maturity we develop mentally, socially, and culturally, instead of continuing to physically grow. Growth during maturity is either obesity or cancer.[21] Therefore, sustainable development is progressive social betterment without growing beyond ecological carrying capacity: achieving human well-being without exceeding the Earth's twin capacities for natural resource regeneration (*e.g.*, trees and water) and waste absorption (*e.g.*, carbon dioxide and toxic chemicals).

Sustainable development is the ability to co-exist in a way that maintains a healthy natural environment, economic well-being, and an equal opportunity for all people on Earth now and in the future. The three are interdependent. Nature is our life-support; it is the life support system for our children and our children's children as well. Only when we have a healthy natural environment, coupled with healthy social systems, can we truly prosper economically.

It is about equal consideration between economic development and environmental quality, between technological innovation and community stability, and between investment in people and investment in infrastructure. Because it is broadly based…cutting across all dimensions of human life, including such issues as energy shortages, species extinctions, pollution, disease, breakdown in families, armed conflict, child abuse, poverty, and corruption…sustainable development requires the participation of the whole of society.

The Unified World View shows sustainability as the ultimate goal. Sustainability could likely hinge upon the acceptance of the fact that everything, including humans and non-humans, are interconnected, interdependent, and interactive. The whole is greater than the sum of its parts. Nature determines the limitations of human endeavors; the integrity of the environment and its ecological processes has primacy over human desires. But when such desires destroy the ecosystem's ability to provide life-support services for future generations nature's dominance becomes apparent in the outcomes we observe (*e.g.* destruction of the ozone layer).

The poor and disenfranchised as well as future generations have rights that must be accounted for in present decisions and actions. And maybe most importantly, the relevancy of knowledge depends on its context in relationship to the circumstances of its receivers.[22] In other words, people must be presented with information relevant to them and their culture, and allowed to fully consider and express what they know.

HITTING LIMITS

Sustainable development is a dynamic process which enables all people to realize their potential and to improve their quality of life in ways which simultaneously protect and enhance our Earth's life-support systems. These are the main poles of tension. Social inequity, the material disparity that accompanies it, and the question of why consideration for nature should come before the welfare of humans, are at the center of the sustainable development debate.

Ecological sustainability is the simple part of the sustainable development concept. While there is considerable debate over where exactly the limits are, there is general consensus that we must learn to live together within the means of nature. Socioeconomic sustainability, however, is a more difficult and potentially contentious concept. The question of "who gets what (and how)" raises the specter of potential conflict both within and between nations. The desire for shared justice and the associated latent conflict is the most scary and politically taxing part of the sustainability question.

Consider all the press now about global climate change and potential resulting dangerous economic and social impacts. If changes in our environment affect our climate, evidence suggests these climate alterations will impact coastal cities with flooding, change the makeup of whole bio-regions around the world, and directly affect the comforts of society. The increasing gap between rich and poor nations, and the growing environmental, cultural, ethnic, religious, and social crises around the world are a primary factor stimulating this awareness. The growing scale of diseases, social and environmental risks, human migrations, high population levels in developing countries, increasing levels of consumerism, violence, terrorism, and intolerance are critical factors alerting decision-makers, market forces, and civil society in general, that the welfare of humanity is at risk. Thus, sustainable development is not merely about a series of technical fixes; it is about re-designing humanity to save it.

*   *   *

Native Americans think about "the seventh generation" and make decisions that would ensure the natural resource returns far into the future, being as stable, diverse, and secure for these next generations of people and their ways of life as possible. In the face of absolute ignorance they would be conservative by following this philosophy. Realizing that they would never know more than a small part of what they need to know, they would keep the scale of their projects small, and be ready to go back when things go sour.[23]

Unfortunately today we tend to think that all the stuff we have is what makes us happy. Our thoughts may turn to our children in what we can leave them as far as money, or stocks, or houses, or possessions but we don't tend to think about the natural resources we will leave them. In this regard, we are not really looking down the road. Many don't give a second thought to aligning overabundance with waste. Or, get the fact that if we spray our yards with a substance that guarantees to kill just about everything, that very substance may indeed poison us or our children as they play in the yard, or drink the water from the well that sits below the ground.

Sustainability is a challenge in honesty. It challenges how we think and why: it looks at common assumptions about economy, society, and nature. Sustainability shows how connected the components of our world are: it gives us back responsibility as to how much respect we show nature's delicate balance and our children's future. And it solicits the question: Does that addiction to more, more, more of whatever the latest trend is really give us peace of mind?

WHAT SUSTAINABILITY IS NOT

Sustainable development is not walking a tight rope, seeking some mythical balance between economics and environment.[24] This leads to habitats half protected, economies weakened, and personal principles bargained away. Instead we must search for ways to create co-action while doing no harm to the life-giving environmental elements that sustain the future of people. Economic activity can promote a healthy environment and healthy ecosystems can enrich their inhabitants.

Also important is the fact that sustainability is not a "thing" but rather is a process: a process that uses common sense and intuition as a baseline. Sustainable development is not strictly a problem of science, engineering, or economics, but is also founded on values, ethics, and the contributions of different cultures.

But, sustainable development is not necessarily popular with the people who can most make a difference by understanding and carrying out its meaning. Problems connecting with sustainable development come from two directions. First, it suggests unwanted sacrifices on the part of individuals craving to preserve the status quo.[25] Secondly, the full unfolding of sustainability involves patience and the ability to look to the future. In this regard, there are often not instantaneous gratifications from actions we might take to fix what's going wrong, thus discouraging further efforts. Immediate solutions are not always apparent to problems people face in dealing with daily struggles. As we get caught up in wanting immediate solutions, we unintentionally end up creating our own demise.

Sustainability is not a trend or phase or even a conditioned pattern. It is not a state in which a compromise (some win; some lose) can be struck. To be sustainable requires ultimate agreement on everybody's part (everybody is a winner). Only partially implementing sustainable development defeats sustainability altogether. Leave one process out of the equation, or in some other way alter a connection between important economic and environmental or social elements, and the system as a whole will gradually be deflected toward an outcome other than that which was originally intended. It is thus critical during the planning of any process involving change to consider carefully the elements of sustainability and the relationships that make them work, if the plan is to succeed.

SUSTAINABILITY'S THREE BASIC ELEMENTS

The process of sustainable development must remain flexible, because what works in one community may not work in another or may work for different reasons. For decisions and actions to be sustainable, they must be ever flexible, adaptable, and creative. Looking at the three basic elements of sustainable development and exploring their meanings in more detail, offers us a better sense for the interconnectedness of these elements and how they provide the foundation for guiding sustainable activities.

ECONOMY: (*Compatible with Nature*)—considering economic development plans that protect and/or enhance natural resource quantities and qualities through improvements in management practices/policies, technology, efficiency, and changes in lifestyle.

ECOLOGY: (*Natural Ecosystem Capacity*)—understanding the natural system processes and integrity of landscapes, watersheds, and aquatic ecosystems to guide design of sound economic development strategies.

EQUITY: (*Balancing the Playing Field*)—achieving total societal welfare by guaranteeing equal access to jobs (income), education, natural resources, and services for all people.

Every project that focuses its efforts in a sustainable context, means we strive to link economic, social, and ecologic parts of the community to strengthen its overall fabric. For example, each element of the overlapping circles diagram below is interconnected (the almond-shaped black area in the middle where the contents of each circle overlap) to demonstrate the interaction between all areas of life—representing equal consideration of all areas of life. To isolate one from the others is not an accurate depiction of the process of sustainable development and the values used to implement it.

Another way of looking at this concept of sustainable development involves considering a three-legged stool, where each leg respectively represents one of the basic elements—economic vitality, ecologic integrity, and social equity. If one of the stool legs is removed, the stool falls over—emphasizing the importance of all three legs to maintaining the upright position of the stool as equally all three elements are important to the satisfying of sustainable development objectives.

Sustainability Model

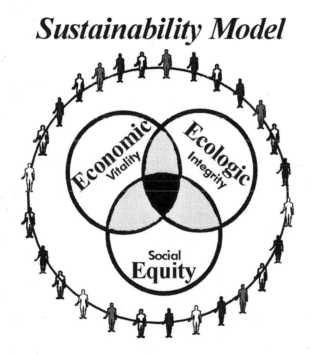

1 *Diagrammatic representation of the sustainability model showing the interconnection among community sectors that should occur in planning projects.*

But, what are the actual objectives of these three circles...Economic vitality, Ecologic integrity, and social Equity? To act in a sustainable

development fashion includes a major transformation in society, focusing on the following:[26]
  • population stabilization
  • efficient, effective use of natural resources
  • refined market economies
  • waste reduction and pollution prevention
  • new technologies and technology transfer
  • "win-win" situations
  • integrated environmental systems management
  • environmental limits definition
  • education
  • perception, attitude and behavioral changes (paradigm shifts)
  • social and cultural changes

It is worth noting that the first issue above involves the growth of human populations. The second, third, and fourth issues involve how humans consume materials and resources. In this regard, to many who work in the arena of sustainable development, there are two primary "Laws of Sustainability." [27] The first law of sustainability states that current population growth and/or growth in rates of consumption of resources cannot be sustained. The second law of sustainability states that the larger the population of a society and/or the larger its rate of consumption of resources, the more difficult it will be to transform the society to a condition of sustainability.

Economy and society are no less important to humanity than ecology, but rather there is a "directionality" of dependence.[28] All life depends on natural resources. In order to account for this directionality, think about the "three-stage rocket ship" propelling or building and enhancing natural capital (ecology) first, human capital (social equity) second, and finally, financial capital (economy). The failure to protect the physical environment threatens the future as well as compromises the present. Once we acknowledge this hierarchical dependence, decisions have a better chance to take us in the direction we want to go.

There is no economy outside of society

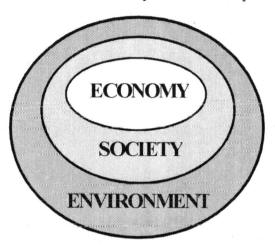

**Socio-economic activities all
happen within the environment**

2 The directionality that guides cross sector thinking when considering the hierarchy of economy, society, and environment issues in evaluating the sustainability of any action or project.

Sustainable development proponents argue that problems in the economy, environment, and society are interrelated and global in context. Economic prosperity can only truly occur alongside a healthy natural environment, coupled with healthy social systems.[29] This is best demonstrated in the hierarchical presentation of the diagram above that illustrates how the socio-economic spheres of communities must always be considered within the larger environmental sphere of influence. This conceptualization suggests how economic and cultural activities are integrated into natural processes.

No human activity can occur without some connections to the environment. In addition, change is the norm, not the exception, and we must take the long-term view, one of thinking far into the future.[30] Also, it is vital to think about how parts of any system of economy, humans, and environment tend to reverse back on themselves without notice (often referred to as a system's feedback loops). This thinking should be directed at both industrial and natural systems, in the context that we can, and should, mimic nature. Pursuing this approach offers flexibility to our thinking, decisions, and actions.

Chapter 2

Our Ecological Footprint

"From a universal standard each moment provides humanity with one
option which, if chosen, will evolve it's consciousness. Any other choices
will result in experiences from which humanity will learn. Eventually
humanity becomes saturated by learning from experiences and is able to
make the choice to learn through enlightenment instead."

Diane Lancaster

In a world where there is a huge gap between how different people
live there is also an enormous difference in how they affect other people around them by the size of their "footprint" on the Earth. Nepal for
example, is a city that is literally being buried by mud slides due to the
deforestation that has occurred. This cutting of forests that held India's
slopes in place is a direct result of corporate appetites—a direct result
of someone's demand for resources from somewhere else. The lands
used by Southern Hemisphere countries to produce food that is
exported for upper income consumers, usually in Northern
Hemisphere countries, affect people these consumers in the north will
never meet.

Between 1960 and 1980 globally, 28.4 million people were displaced,
many left desolate due to their small farms being over-run by corporate food production. More recently, in India 20 million people have

been left desolate due to environmental damage from corporate development projects to better support the needs of growing affluent societies elsewhere. Likewise, in Thailand 10 million face eviction to make way for tree farms.[31]

All of this displacement is due to higher world demands for the many resources that can be produced in greater quantities by massive industrial production. Thus, because of someone's footprint (demand) somewhere else, this person has a quality of life that far exceeds that of the poor, environmental refugee. Someone far away ultimately affects a people who once lived off of the land now providing resources to that far-away consumer. And once again consumer-needs have indirectly displaced a culture, leaving the poor even poorer.

Impacts from a person's actions or footprint in a place are also represented in their contributions of waste to the environment. The predicted impacts of global climate change from waste gases produced by automobiles and industry, will probably exacerbate hunger and poverty around the world. People who are highly dependent upon farming, fishing, or forestry will see their livelihoods destroyed.[32] The poor will suffer the most because they have fewer options for responding to climate change impacts on our environment.

Unfortunately, many who consume the majority of the Earth's bounty, directly or indirectly, have no idea the path of these products or the severe imbalance and destruction that these products have left in their travels on the way to the consumer. Good intentions aside, the information consumers receive is more than slanted about how their needs are being met. Are these the actions of responsible, world citizens?

CONSUMPTION

As we begin to look at an individual's impact on our world it is vital to look at the reality of what the Earth can provide. If you divide all the "good" land on Earth (excluding places not able to support prolonged human existence) by the present human population of 6 billion people, there are only about 3.7 acres of bio-productive space for every person in the world to meet their needs. This does not include 25 million other species on Earth who also must meet their needs. The United Nations states that the 1.2 billion people around the world who live in afflu-ence, like many Americans, consume over three-quarters of the Earth's total output. The remaining 4.8 billion people—80 percent of the pop-ulation—survive on less than a quarter of world output.[33] Thus, everyone on Earth will never be able to reach the lifestyle enjoyed by most Americans.

Likewise, the U.S. has less than 5 percent of the world's population, yet we use 25 to 35 percent of the world's resources and produce 25 percent of the world's waste.[34] Americans produce 21 percent of all goods and services. The average U.S. citizen, when compared to the average citizen of India, uses:
- 50 times more steel
- 56 times more energy
- 170 times more synthetic rubber
- 250 times more motor fuel
- 300 times more plastic

On less of a global scale it is even more staggering. One (1) percent of the U.S.'s population owns about 80 percent of the country's prop-erty and controls 90 percent of its wealth. Will laws restricting popula-tion growth and resource consumption really affect these few people in the U.S. or the populations of the more industrialized nations?

And, what about up and coming more populous countries of the world? Under-developed countries have begun developing better lifestyles. With this shift, the burden on global resources will only be magnified. China and India, for example, have the potential to more than triple what America already consumes. In addition, these very populous countries have the capacity to undo gains that we might have already made under some global agreements, such as climate change.

If each Chinese person were to consume just one extra chicken per year and if that chicken were to be raised primarily on grain, this would require as much additional grain production as all the grain exports of Canada, the second largest exporter.[35] If China's per capita consumption of beef, currently less than 2 lbs per year, were to match the U.S.'s approximately 20 lbs per year, this would require additional grain production equivalent to the entire U.S. harvest, less than one third of which is exported. If China were to consume seafood at Japan's per capita rate, it would need more than 100 million tons, more than today's total world catch. If the Chinese were to use wood products at the Japanese rate, their demand would exceed Japan's more than nine times. But, can industrialized countries really ask China and India to remain forever at a greatly lower economic level, than for example the U.S.?

Everyone of us has an impact on our planet. This is not bad as long as we don't take more from the Earth than it has to offer. In the usage of timber alone the people in the U.S. should cut consumption by 79% to bring timber use into a more sustainable arena, in relation to the rest of potential world consumers. The Netherlands would have to reduce its timber use by 60%. And this is only one small piece of what we take from nature.

Unlike consumption by other animals, consumption by people is not determined solely by biology. While most species on Earth consume little beyond their food, the bulk of human material consumption consists of manufactured, non-food items such as energy, clothing, automobiles, and a vast array of other goods, often far away from where we live.[36] Therefore, the load imposed on the world by our biological metabolism is vastly increased by our industrial and technical metabolism. Likewise, impacts of waste products going into the environment are felt much further away than those simple waste products from a herd of deer or pod of whales. The First Nation peoples in northern Canada lost their forests and livelihood from acid rain, originally produced thousands of miles away by the wastes of someone else.

HOW BIG IS YOUR FOOTPRINT?

As societal living standards rise, more and more people will live on the ecological carrying capacity "imported" from somewhere else. Why? Because increased resource consumption and waste assimilation requirements of a defined human population or economy represent a corresponding increased need of productive land area. And the land area they live directly on can only produce so many resources and provide so much waste assimilation capacity. When limits on these functions of nature are reached, these functions must be served by lands somewhere else.

People who live on ecological goods imported from afar are also spatially and psychologically disconnected from the resources that sustain them. Thus, they lose any direct incentive to conserve their own local resources and have no hand in the management of the distant sources of supply. There may well be greater ecological, community, and personal merit in learning to live more simply so others can live at all. Consideration of our ecological footprint equates to carrying on a responsible, sustainable lifestyle.

The vital question then becomes: How far from a person's own home should they truly and rightly reach for the resources from someplace else? How far away should one place the waste of materials? Responsibility falls to both venues. People live in that someplace else. People may need that deported resource. The ability to use another's resources to maintain our lifestyle has shown that it only leaves destruction and refugees in its path. The generations that have been born into the era of the consumer have before them an opportunity to show consideration for other life.

Consider for example, the ways of big city life. This lifestyle breaks natural material cycles and provides little sense of our intimate connection with nature.[37] New York City cannot grow its own food and must import it from elsewhere. Trade, thus leads to a situation in which every city and nation is trying to live beyond its own absorptive and regenerative capacities by importing these capacities from elsewhere. Since many of us spend our lives in cities and consume goods imported from all over the world, we tend to experience nature merely as a collection of commodities or a place for recreation, rather than as the very source of our lives and well-being.

The first step toward reducing our ecological impact is to recognize that the "environmental crisis" is less an environmental and technical problem than it is a human behavioral and social one. It is imperative to truly recognize how far an individual's consumer demands really reach. For example, it has been calculated that the head of lettuce you purchase in your local supermarket has traveled an average of 1,200 miles from where it was grown.[38]

There is also the story about two tomato hauling trucks passing each other in the Sacramento Valley (California), one entering the Valley and the other leaving. The tomato truck coming into the Valley was carrying tomatoes grown in Vera Cruz, Mexico. The truck of tomatoes leaving the Sacramento Valley was traveling with its contents to San Francisco, to then be shipped to a tomato processing plant in a small Mexican town just over the Border. What is wrong with this picture?

When we walk on powdery snow or run along a sand beach we leave our footprints as record of our being there. When we step into a puddle of water our footprint impact is reflected in the ripples and waves created in this puddle, that reach well beyond the depression of our step. Likewise, our social footprint is the number of people we interact with in our community and beyond. Playing excessively loud music in the middle of the night for example, having an impact on our neighbors, is one measure of our social footprint.

Wackernagel and Rees call consumption of goods and services from our environment our "ecological footprint".[39] This ecological footprint measures flows of energy and materials from nature to and from a region's economy. Our actual footprint is then converted to quantify and illustrate the corresponding land/water area required from nature to support these flows, often hundreds of miles away from where we live.

The ecological footprint measures what we do to nature—how our way of life affects nature. Ecological footprint auditing is like an accounting tool. It helps us look at how much of nature we are using up as well as how much of our waste nature has to soak up. This is not bad as long as we don't take more from the Earth than it has to offer. The ecological footprint is able to reveal the lack of sustainability present and the imbalances of society that will affect future generations.

WASTE

We as Americans consume 120 pounds of stuff daily and generate at least 4 pounds of trash daily per person. This does not include hazardous waste, municipal waste, industrial waste, and solid waste. When all sources of waste are considered, it is estimated that more than 50 tons per person per year is produced.[40] When multiplied by the U.S. population, we have to ask: where is all this waste going?

How much carbon monoxide does the average automobile send into the air for every hundred miles driven? If you left your automobile at home a couple of times a week, and walked, bicycled, or car pooled, you would reduce CO_2 emissions by 10,000 pounds per year. Then multiply this by all the drivers on the road or the 508 million cars in the total car fleet of the world.[41]

Approximately two-thirds of residential water use is for toilet flushing and bathing. A 20 minute shower uses 15 gallons of water; an average bath uses 30-50 gallons of water; flushing the toilet uses 4 gallons of water; a dishwasher uses 8-12 gallons of water; and a top-loading clothes washer uses 40-55 gallons of water.[42] How much water do you use daily in these activities? Do the math and multiply it by the number of people on the Earth, 6 billion as of October 1999.

The average quarter-pound hamburger requires 616 gallons of water to create its meat, the cheese requires 56 gallons, and the making of the bun 25 gallons of water.[43] A family of 5 people consume 10 pounds of meat per week, and throw out 2 pounds of food per week.

Now multiply these details by the 270 million in the U.S. population. If only half made small changes imagine the difference.

The demand of our ecological footprint also uses the global ecosphere as a receptacle for local wastes. Think about heating your home with fuel oil. How many toxins does this put into the air daily? How many watts of electricity do you use a day? What pollutant emissions at your local power plant does this create?

Remember the Chernobyl accident in Russia when radioactive waste spread over most of the globe and heavily contaminated cows milk in Scandinavian countries? The impacts of society's ecological footprint can be a silent intruder unforetold and unseen by no fault of our own.

FORESTS

According to an Associated Press report in April 1999, researchers discovered that the destruction of the Brazilian rain forest is accelerating at twice the rate as previously believed. And, the United States, along with other industrialized countries, are the main culprits.

On a global scale, according to the World Commission on Forests and Sustainability, from Siberia to Haiti, the world's forests are presently being exploited far beyond their ability to reproduce. Today, virtually the only economic value officially assigned to forests is related to the role they play in supplying the world with timber and fiber.[44] Nearly 75% of West Africa's tropical forests have been lost since 1950. Thailand lost a third of its forests in just 10 years during the 1980s. Forests face an even shakier future with the global population expected to grow 50% in the next half century.

The loss of millions of acres of forest cover from the Earth every year is especially serious because of the ecological services forests provide to all forms of life, but in particular humans.[45] Forests enhance the hydrological cycle, protect against soil erosion, offer habitats to enhance biological diversity, and even control global weather patterns.

While the economic products of forests (timber & fiber) can be partially substituted for, the forest's ecological services in support of a well-functioning world cannot.

3 Diverse plant assemblages and massive root system of a tree in a Jamaica rainforest.

This is especially evident in the long troubled Chesapeake Bay ecosystem of the U.S. Evidence of extensive farmland and other rural development activities, causing excessive nutrient runoff to the Bay, has been accumulating for years. Much of the watershed development around this massive Bay has been related to the cutting of forests. Tree canopy in the Chesapeake watershed has declined from 51 percent cover to 37 percent in the last 25 years.[46] These trees are one of the main players in decreasing nutrient-runoff to the Bay and directly affecting water pollution patterns.

We can satisfy the world's material needs from forests without jeopardizing their ecological services. But, consideration for the importance of forests must go beyond their literal and traditional economic value and account for their being the largest reservoir for plants and animals on land, their role in maintaining supplies of clean water, in creating and retaining soil, in contributing to the productivity of fisheries and agriculture, and even helping to regulate climate.

And very importantly, at this moment when we have never been better poised technologically to evaluate natural products from rain forest plants and animals for medicines and other valuable products, the species themselves are disappearing. But, issues of survival in many developing countries, with pressure for loan repayments to industrialized countries, cause poor people to look at the most immediate way of relieving their debts...cut their forests for the profit or the promised income of cleared, agricultural land.

For example, soon after the discovery of prostratin, an active drug against the HIV virus produced from the bark of the masuala tree, loggers arrived in western Samoa to clear-cut the rain forest where this tree grows. Required by the government to pay for a new school, villagers in Falealupo, one of the villages in this forest, initially believed they had no recourse but to accept the loggers offer of $1.83 per acre for their 30,000 acre rain forest. Fortunately after much soul-searching, village chiefs were finally able to bring together the resources from outside donors to pay for the school and spare the forest.[47]

Chapter 3

Beyond the Obvious

One who stands at the edge of a cliff is wise to define progress as a step backwards.

As a whole, everyone always does the best they can in any given circumstance: from the most vile murderer on death row to the President in Washington D.C.; from the welfare mom in the projects, to the millionaire on Wall Street; from the leader of a street gang in the ghetto, to the Pope in Vatican City. These folks are doing the best they can. Yet unfortunately our track record is a bit scary. There is example after example after example of how great governments have ultimately self-destructed: lost in their own grandiosity; blinded by their own blazing sword; continuing to believe that destruction of the very lands that nurture them constituted some form of their cultural growth.

For example, the early Sumerians were a great civilization. One of their first kings was a man named Gilgamesh. He was the first mortal who decided to defy a God. He opposed their forest God, Humbaba. Humbaba was the God that had been entrusted to protect the cedar forests of Lebanon from mankind. King Gilgamesh wanted to build a great city in his own honor, as he was a King. So King Gilgamesh and his loggers began to cut the forests that stretched from Jordan to the sea of Lebanon, taking this region from 90% forests to 7% forests in just 1,500 years. This in turn led to a reduction in downwind rainfall of

80% because of the drastic loss of trees, as a tree's root system is critical to the water cycle.[48]

Ultimately King Gilgamesh decapitated Humbaba. Thus, the main God, Enlil avenged Humbaba's death by making the water undrinkable and the fields barren. King Gilgamesh and all his people died.

The legend of Gilgamesh is one of the earliest recorded examples of a civilization being wiped out as a result of its extreme destruction of natural resources. At the same time, Mesopotamia was becoming overpopulated due to two major factors: the continual conquering of neighboring lands so as to commandeer their forests and the natural consequence of changing from a nomadic to a communal lifestyle. In order to feed the expanding population, the irrigated land was exhausted by the continual planting and harvesting of barley. The end result was the once lush, green land of Lebanon was turned into a desert. And, it remains that way today.[49] This destruction caused the extinction of the entire Sumerian civilization.[50]

EXTINCTION REPEATED

"There is honor in looking back and respecting the past."

Steven Spielberg,
1999 Academy Awards Presentation

The Greeks came after the Sumerians, and they followed the same pattern with the same results. The Romans then followed the Greeks. The same mistakes were made and the same consequence occurred; mighty empires collapsed. Every major civilization of the last 7,000 years has self-destructed. And, their destruction was caused by a shortage in their primary fuel supply due to deforestation, the destruction of its watershed, and the resulting soil and water depletion.

Gretchen Daily's description of past cultures on south Pacific Islands offer additional insight with regards to many historical accounts of how human interaction with the environment, and eventually with each other, played out.[51] The many Polynesian Island histories shared very similar initial conditions, with small groups of people of similar cultural origins (from Polynesia) colonizing the Islands, introducing to each Island the similar basic agricultural practices. Yet, what happened from there varied tremendously.

The most interesting story occurred on Easter Island. Original inhabitants, who colonized it 1,500 years ago, apparently lived well. They ate meat from porpoises and hunted from canoes they fashioned from giant palm trees. Before long, however, the forest was cleared, inducing soil erosion and drying of streams, along with an abrupt halt to the hunting and cooking of porpoise meat...both the canoes and fuel wood were derived from the forest. Island inhabitants next wiped out coastal marine and bird food resources, then resorted to eating rats, and finally turned on one another. Because of the onset of cannibalism, population numbers quickly decreased from somewhere over 10,000 at its peak, to only about 2,000 people when the Island was discovered by Europeans in 1722.[52]

Easter Island was not the only story of this type in the Pacific Island chain to suggest unchecked population growth, destruction of natural resources, social terror, and cannibalism. But there did exist stories that stand in stark contrast to the ruthless, survivalistic manners of such entities as the residents on Easter Island. Polynesian societies on other islands also underwent an initial period of environmental degradation, but then, in stark contrast, developed sustainable economies that have persisted through today in relative peace and comfort.

What caused the divergence in environmental and social trajectories? Kirch attributes it to "conscious choices,"[53] documented in oral history and archeological remains. These choices would not have been easy, nor would their enforcement have been pleasant. They involved eliminating environmentally degrading animal culture from the agricultural production system and imposing rigid population control measures, such as celibacy, abortion, infanticide, and even expulsion of some of the population. Patterns of sustainable populations may have also been influenced by island size because evidence suggests populations on the tiniest of islands were more likely to persist.

Two issues from this history of Pacific Island cultures stand out as relevant to present global circumstances according to Daily.[54] First, scientific understanding of human interactions with the environment is necessary, but insufficient to prevent irreversible destruction of life support systems. We, presumably like the islanders cutting down the last trees on Easter Island, know more than enough scientifically to recognize trouble and start moving in the right direction. But uncertainties are capitalized on by our social structure to head-off logical actions based upon our knowledge of history. Clearly, most of the required action is on the social side rather than the science side.

Second, human beings are much more comfortable interacting in small regional/local settings. Our future prospects seem to depend on whether we can foster enough courage and foresight to recognize the global magnitude of our impacts on environments and people to create the economic, legal, and other social institutions needed to bring human impacts into balance with what the entire globe can sustain. Again, the action is on the social side. This is why efforts on the part of scientists to foster dialogue on environmental issues are so important to have with everyday people as well as leaders in the social sciences, business, and government.

ACCELERATING CYCLES

It is taking less and less time for civilizations to collapse. The Sumerians took approximately 4,000 years to self-destruct; the Greeks 1,400 years; the Romans only 1,100 years. Now, the United States is walking the same road. And, we are barely 200 years old. At the present rate of consumption, the U.S. will be stripped almost bare of its forests in 50 years.[55]

Humans and the natural world are on a collision course. Change is occurring faster and faster, and the acceleration continues.

- It took four million years of human history for the global population to reach its first billion, in 1804.

- In 1927, 123 years later, we had passed two billion people on Earth.
- By 1960, 33 years later, the population had reached three billion.
- By 1974 four billion.
- By 1987 five billion.
- Now, into the 21st century, we passed six billion people on Earth in late 1999.
- It only takes a dozen years to add a billion.[56]

Someone once figured out how long it would take to count to a billion. If you did nothing else except count one number every half second, twelve hours a day...it would take 32 years, give or take a few months.

If current predictions of population growth prove accurate and patterns of human activity on the planet remain unchanged, science and technology may not be able to prevent either irreversible degradation of the environment or continued poverty for much of the world. The Earth's ability to provide for growing numbers of people is finite, and we are fast approaching many of Earth's limits. Pressures resulting from unrestrained population growth put demands on the natural world that can overwhelm any efforts to achieve a sustainable future.[57]

Each year now, scientists report new evidence that the human economy has exceeded the ecosystem's limits for providing natural resources as well as its abilities to serve as receptacles for waste products. Basically humanity consumes more than nature can provide and creates more waste than can be recycled. From advances in environmental sciences, satellite remote sensing, and other monitoring systems, a more comprehensive assessment of local and global environmental deterioration has now become possible.[58] In summary, evidence is accumulating with respect to accelerating

- loss of vital rain forests,
- species extinction,
- depletion of ocean fisheries,

- shortages of fresh water in some areas and increased flooding in others,
- soil erosion,
- depletion and pollution of underground aquifers that supply drinking water,
- decreases in quantity of irrigation water,
- growing global pollution of the atmosphere and oceans, even in the polar regions, and
- ice volume decreases from the glaciers in Greenland over the last 5 years, apparently fueled by warming global temperature, which may contribute to sea level rise.

For example, nature creates new topsoil each year, but in much of the world (particularly in the U.S.) we destroy topsoil faster than nature can create it. In the past two hundred years we have used up 75% of our topsoil.[59] And, the amount of loss is accelerating. Loss of topsoil reduces our future farming capabilities. Topsoil destroyed is topsoil taken from our children and grandchildren. Last year we lost 7,000,000,000 tons of topsoil alone.[60] And, our supply of good water is disappearing at the same terrifying rate.[61]

When our nation began its ascent to its present status as the most wealthy and powerful nation on Earth, wood was our main fuel. With the discovery of oil in Pennsylvania in the late 1850s, we set in motion what has resulted in the U.S. being the grocery store to the world—and, in return, has led most of the world to be dependent on all that is produced by oil.[62] At present, estimates are that there is enough oil in the world's reserves to only last for the next 45 years at our current rate of consumption. Can you imagine gas prices as fuel becomes more and more precious?[63] It will make the 1970s look like a walk in the park. And if the start of the new century is any indication, we are facing constantly increasing fuel costs.

Futurists believe there will be food enough for only three (3) billion people by 2050. Seven (7) billion people will be facing starvation. The most widespread hunger ever. Exploding populations colliding with short food supplies has already led to mass starvation in places such as Haiti, portions of the Far East, and many of the nations of Africa.

With so many nations of the world in dire straits it may help to look at this from the global perspective. U.S. involvement in World War II was partially initiated by the Japanese attack on Pearl Harbor. We now know that attack was due in large part to the United States cutting off the supply of mid-eastern oil to Japan with a naval blockade. Now we hear stories of our government passing nuclear secrets to China, the most populous nation on Earth. A nation projected to begin importing 10 million tons of grain per year, demanding more food and other goods than any other nation in history. The potential for demanding control of the situation is enormous—we have the food; China has nuclear power and is starving. Who do you think will back down?

SUCH A FINE LINE

The story of Gilgamesh is a tale of ancient folklore. Folklore passes history on and tells morality tales about their deeds and the deeds performed by prior generations to the generations to come. When our children and grandchildren write their folklore about us, what will they say? Will it read that we were remembered as the generation that could have brought the Earth back from the brink of destruction, but failed to act? That we sold out their futures for our present luxuries?

History is relentless, preserving, persistent, and never looks back. There is a fine line between comfort and luxury. Luxuries in and of themselves are not necessarily bad things. But, luxury has a price: a price that is paid for by the destruction of the environment; paid for by the exploitation of cheap child labor; paid for by investments in ecologically harmful companies; and consequently paid for by more than just one's own personal dollars in striving to obtain these luxuries. It is then that they cease to be luxuries and commence being decadent indulgences.

Globally, the promotion of rampant consumerism is linking the individual's self-esteem and ability to find happiness with how much "stuff" is accumulated. From where we live to what we drive to how we dress has become the barometer used to gauge worth. We are judged by these tokens, manipulated by them, and now are feeling the consequences of their excess.

There was a time in human history when we could go someplace else. Our planet Earth is no longer empty; it is full. Economic growth has successfully provided goods, services, and jobs, although less than 20% of the world population benefits from this prosperity.[64] But now the economy has grown so large that despite best technology, both natural resources to support our needs and places to put our wastes (the environment), are becoming scarce on a global scale for the first time ever.

Reclaiming sustainable communities means that we must re-learn much of what we have forgotten. We must remember and reconstruct the cultural values we once had regarding the human community's place in nature, but lost sight of. People can harbor such a deep longing for stable, sustainable communities that they are totally committed to practicing sustainability, which is based first and foremost on meaningful interpersonal relationships that promote a connection between humans and their natural world, as our forefathers understood. Only then will recreating a community's physical attributes have any significance toward a more sustainable society. This is one way we might repair our social fabric, insulating us from actions that degraded the environment in past generations.

As the new ideal of sustainability emerges, there will be renewed interest in understanding the methods and uncovering the wisdom of traditional resource uses from past cultures. Modern day society needs to find out what worked sustainably in the past and begin recreating it in the present. Where problems arise, work together to resolve them. For in some ancient civilizations are examples of human resource use that did not profoundly interrupt cycles of the natural world and its constant renewal. The only way to create, maintain, and pass forward the sense of community, including its harmony with the natural world, is founded on the quality of interpersonal relationships [65] and what we can learn from the past experiences of previous cultures and our forefathers.

Section II

The Devil is in the Details

Chapter 4

The Chemical Story

"Sentiment without action is the ruin of the soul"

Edward Abbey

Chemicals in our world originate from two sources. First, chemicals are found naturally in our Earth's crust. Life on Earth has evolved over millions of years to accept small levels of these natural occurring materials. Animal and plant cells, however, don't know how to handle large amounts of lead, mercury, radioactive materials, and other hazardous compounds from our Earth's crust. Secondly, society also relies on thousands of manufactured chemicals. Since World War II, we have produced more than 70,000 chemicals, such as dichloro diphenyl trichloroethane (DDT), a pesticide, and polychorinated biphenyls (PCBs). American homeowners used 32 million pounds of pesticides on their lawns in 1994. Household gardeners use pesticides at 10-20 times the rate farmers do. Many do not go away, but rather, spread and accumulate in the fat cells of animals and humans. These chemicals often build up in living tissue, leading to cancer, hormone disruption, and development defects.[66]

A HISTORY

The production and release of vast quantities of novel synthetic chemicals over the past 75 years has proved to be a great global experiment, one that now involves all life. Pesticide use is a good example of our producing wastes faster than nature can absorb them. If nature could absorb pesticide residues as fast as we created these substances, then there would be no buildup of toxic residues and we could not regularly measure them in the environment and animals, as we now do.

But there has been a measurable buildup of pesticides, even in such remote places as the north and south poles, where these synthetic chemicals have never been used. They are also found at the bottom of the deepest oceans, in the drinking water of much of the midwestern U.S., and in the breast milk of women worldwide.[67] We have clearly exceeded nature's capacity to absorb pesticide wastes.

Even before the Chemical Revolution moved into high gear at the end of World War II, the first warning sign appeared that some man-made chemicals might spell serious trouble. In 1944, scientists found residues of a man-made pesticide, DDT, in human fat. Seven years later, another study brought disturbing news of DDT contamination in the milk of nursing mothers. In the early 1950s, naturalists saw thinning eggshells and crashing populations of bald eagles and other birds. By 1962, Rachel Carson documented the growing burden of contamination in *Silent Spring*,[68] which detailed the devastating impact of persistent pesticides on wildlife and warned about hazards to human health. More recently zooplankton, small insect-like animals near the bottom of marine and freshwater food chains, have been observed to develop tumors, just as those found in higher life forms.

Ironically, many chemicals that were developed to control disease, increase food production, and improve our standard of living are, in fact, a threat to biodiversity and human health. Who would have thought 30 years ago that chlorinated fluorocarbons (CFCs) used for refrigeration, vastly improving our quality of life, would now be damaging the atmospheric ozone layer, important for the future of all life on Earth?

How about something more recent? At the most basic level, computer chipmakers take beach sand, mold it into silicon wafers and add electronic circuitry to produce semiconductors, the brains behind our computers. The semiconductor industry has propelled Silicon Valley and Wall Street to vast riches. But after years of hailing themselves as the cleanest manufacturing business around, chipmakers now face increased pressure over the health of their workers.[69] Computer manufacturing workers are now alleging that years of exposure to toxic chemicals, including solvents, cause cancer and birth defects. Because the risk from these originally well-intentioned chemicals outweighs their benefits, their continued use is no longer warranted. Today, the contamination from persistent man-made chemicals is a pervasive global problem.

PERSISTENCE IS EVERYTHING

A major health study conducted by the World Health Organization several years ago[70] suggested that chronic conditions and new infectious diseases from lack of our resistance to bacteria, not "dreaded events" such as AIDS and the Ebola virus, were going to be the scourge of the next quarter century. It was the persistence and universal occurrence of chemicals in our environment and their unpredictable, often synergistic, affects on plant, animal, and human life, the report authors had in mind.

It is commonly being observed along our seacoasts that stranded, dying marine mammals, such as dolphins and whales are being killed by viruses. Some scientists, however, believe pollutants are more likely the underlying cause of those deaths.[71] PCBs and other persistent organic pollutants (POPs) are known to suppress the immune system in animals, including humans, making it tougher for them to fight off infectious disease agents like viruses and bacteria.

Because of their unique properties, chemicals classified as POPs pose a special kind of challenge that makes it impossible for any nation to remedy the problem by acting alone. POPs don't degrade readily and, even more important, they don't stay put. They can travel thousands of miles in complex journeys on air, water currents, and through the food web, making any one country's contamination inevitably the world's problem. POPs are now universal.

Even in wild and remote places, POPs are present due to their ability to globe-hop via atmospheric winds and rain, but the contaminants must settle somewhere and come to rest even in locales where man rarely treads. This is because most POPs share a notable physical and chemical characteristic that makes them highly mobile and capable of traveling to the ends of the Earth.[72] These compounds are semi-volatile, a property that allows them to occur either as a solid or a vapor depending on the temperature. Once a persistent contaminant has evaporated in higher temperatures, it can travel great distances in air masses, often hitchhiking on particles in the atmosphere like dust.

Through a process known as the "grasshopper effect," persistent chemicals also can jump around, evaporating in warm conditions and then settling in cool spots. When the temperature is right, POPs will again take flight and continue hop-scotching travels that carry them anywhere and everywhere on Earth. Scientists detect them wherever they look in the world, even in regions where these synthetic chemicals have never been used. The pesticide toxaphene now contaminates fish in wilderness lakes in the Canadian Arctic, but there are no records of its use anywhere near that region. Persistent contaminants typical of industrial regions like the Great Lakes have been found in albatrosses on remote Midway Island in the middle of the Pacific.[73] The penguins in Antarctica have become contaminated with a breakdown product of the pesticide chlordane and other persistent chemicals. There is no clean, uncontaminated place anywhere on Earth and no creature untouched by this chemical legacy.

POPs jeopardize human and wildlife health in all parts of the world. In the tropics continued use of persistent pesticides continue to feed this toxic problem. These toxic chemicals contaminate industrial regions through the release of persistent combustion and manufacturing by-products. In many regions leaking stockpiles are also a source of contamination.

Acute exposure in tropical agriculture has caused large numbers of human deaths and injuries, including severe nervous system and liver damage.[74] Numerous studies have also linked these synthetic chemicals to cancer and other significant health problems in people and wildlife.[75] Emerging science has recently heightened concern about typical "background" levels of these contaminants and a new kind of hazard known as "endocrine disruption." Findings of endocrine disruption in humans and other mammals mean that toxic chemicals are able to mimic natural body chemicals when these pollutants occur in large enough concentrations, for example in humans. This happens by a process of chirality, a characteristic exhibited by chemicals with asymmetric molecules.

This asymmetry causes molecules of the same chemical to exist as mirror images of one another. Many of the building blocks of living things, such as sugars, amino acids, and proteins, are also chiral, like many of the POPs. Thus, the chiral pollutants can serve as mirror images of natural tissue chemicals, tricking the body into substituting the toxic pollutants for the real thing.[76] By this process of acting as replacements in the body, researchers find that PCBs, and their co-contaminants, can do damage at extraordinarily low doses, measured in parts per trillion.

Studies by Case Western Reserve University and University Hospitals of Cleveland Ohio (USA) also suggest a link between exposure to lead on the job and risk of developing Alzheimer's disease, and a connection between use of pesticides in the home and risk of developing Parkinson's disease in adults.[77] Researchers present compelling evidence that environmental toxins can permanently affect nervous system function. The studies provide further evidence that some neurological disorders, long associated with the aging process, may have more to do with a person's work or leisure history than with a person's age.

In recent decades, the incidence of cancer of the testicles in men under age 34 has been increasing rapidly in many countries. Likewise, during the past five years, medical researchers' have published reports of dramatic declines in human male sperm counts and increasing sperm abnormalities that have occurred over the past half century.[78] These reports have caused a contentious debate about whether these changes are, indeed, real. Two of Europe's leading reproductive researchers have hypothesized that increasing exposure to environmental estrogens (several POPs), also included in the endocrine disrupter category, is likely to be responsible not only for lowered sperm counts, but also for genital defects, testicular cancer, and other male reproductive abnormalities.

The semen quality and sperm counts of young Danish men, all of them who must present themselves for medical examinations to determine military fitness, were studied from 1996-98.[79] Results have shown that a high number of men displayed lower than expected sperm counts which were related to decreased fertility. Commenting on this study, the World Wide Fund for Nature (WWF) believes that exposure to endocrine disrupter chemicals in the environment may be a significant factor affecting human fertility.

THREATS TO FUTURE GENERATIONS

Each of us now carries several hundred synthetic chemicals that were not present in the bodies of our great grandparents at the turn of the century. Every child born today has been exposed to persistent chemicals in the womb. Some of the foods most craved by pregnant women, such as meat and dairy products, may give their babies an extra dose of toxic pollutants.[80] Because these chemicals also become concentrated in breast milk due to their affinity for fatty substances, a baby can experience the heaviest exposure to contaminants in its lifetime through breast feeding.[81]

Scientific literature is filled with references to the serious affects that pesticide and other POP poisoning is having on humans, especially our children. Millions of children living in the world's largest cities, particularly in developing countries, are under daily threats of life-threatening air pollution, usually 2-8 times above the World Health Organization's maximum guidelines.[82] Because children are smaller and their organs still developing, they are more vulnerable to air pollution. Cities outside the less developed world are not immune, with New York, Paris, Tokyo, and Los Angeles having among the highest levels of traffic-based pollution in the world. Given the projected growth of cities and the relatively average young age of their populations, efforts to control air pollution are of increasing importance.

4 *The location of this cemetery adjacent to a chemical manufacturing plant along the Mississippi River in Louisiana is ironic with regards to the risks that residents face living in this region, often referred to as "Cancer Alley."*

Studies have now also concluded that children living near heavily traveled streets or highways are at greater risk of developing cancer, as measured for example in Denver, CO (USA).[83] Homes adjacent to street corridors carrying 20,000 or more vehicles per day had roughly a six-fold increase in risk for children contracting cancer, including leukemia. Motor vehicles are a significant source of air pollution emissions, including benzene and other organic compounds. Occupational exposure to elevated levels of benzene is a known cause of leukemia in adults. Not only direct air exposure seems to present problems, but even exposure to the neighborhood soils during play where benzene and other emission chemicals have been shown to accumulate, suggest high risks. Likewise, a study in Stockholm, Sweden several years ago showed correlations between nitrogen oxide pollution from motor vehicles and cancer rates of nearby residents.

This exposure by humans to chemicals in the environment threatens the integrity of the next generation. Consider the following. A team of U.S. and Canadian researchers recently documented that the amount of pesticides and other man-made chemicals in the amniotic fluid of unborn babies was high enough to disrupt the natural fetal hormones.[84] These chemicals are observed to interfere with the action of the male hormone testosterone interfering with the normal effects of testosterone on the growing fetus. Given these immense stakes, precaution dictates swift and strong action to eliminate the use and production of persistent chemicals, because POPs, by their nature, cannot be managed.

A new U.S. report by the Clean Water Fund and the Physicians for Social Justice [85] links chemicals used by industry and at home to human developmental disabilities. These chemicals were found to be toxic to the infant developing brain and can lead to hyperactivity, attention deficit, lower IQ, and motor skill impairment. The report found that blood levels of lead in one million U.S. children now exceed the accepted level above which lead affects behavior and cognition. The report also found over 80% of adults and 90% of U.S. children have residues of one or more harmful pesticides in their bodies.

A United Kingdom review of available evidence stated environmental pollution from lead, PCBs, and radiation is harming the intelligence of millions of people.[86] The blood level of lead in 1 of 10 UK children was found to be high enough for intelligence to be affected, while the same was true of 9 in 10 children in many African cities. The intelligence of Inuit children in the Arctic is also being damaged by PCBs that originate in the tropics and arrive in northern Canada within a week.

Our standards for safeguarding drinking water for our children aren't keeping up with the chemicals being produced. After accidents, cancer is the largest killer of children in the U.S., and cancer in children is on the rise. Is there an environmental connection here? Brain, nervous system cancers, and acute lymphocytic leukemia represent the majority of cancers attacking children. Large groups of these cases have been associated with regions having drinking water problems (water contaminated with volatile organic compounds) discharged by industry and municipalities into underground sources of drinking water. These case clusters have shown up in places like Toms River (NJ), Winona (TX), Port St. Lucie (FL), and Woburn, (MA). Children cancer clusters are also being investigated in Rochester (NY), Christian County (IL), and McFarland (CA). Do we continue to wait for isolated outbreaks to keep occurring and growing, or do we act now on the presumption that some chemicals in drinking water cause child cancer based on tests in animals?

And then there are those chemicals that have been suspected and are now proving to be even worse than originally thought. The U.S. EPA concludes now that many Americans may have enough dioxin in their bodies to trigger such subtle harmful effects as developmental delays and hormonal changes in men.[87] Dioxins are chlorinated chemicals produced mainly by incinerators and paper bleaching processes. They accumulate in the food chain, winding up in body fat when people eat contaminated animal products. Although the risks of dioxin were recognized a number of years ago, new data evaluations on effects are finding that the risk of getting cancer from dioxin is 10 times higher than previously thought.

STABILITY AND MAGNIFICATION

POPs are highly stable compounds that can accumulate and remain in the environment or in body tissue for years or decades before breaking down. Chemicals characterized as "persistent" resist the natural processes of degeneration by light, chemical reactions, or biological processes that would otherwise eventually render them harmless, as is the case with many natural chemicals. Sometimes, as with the pesticide DDT, the breakdown products that are also toxic, notably DDE, prove far more stable and persistent than the original pesticide.

The body cannot readily excrete persistent contaminants except through breast feeding, so most of the known POPs typically have long half lives in the body. With continued exposure their concentrations in body cells only grow higher over time. Persistent contaminants are now pervasive in the food web, with the animal products of meat, fish, and milk, in particular being the primary routes of human exposure because of their fatty content.

We are also now learning about the co-occurring reactions that many chemicals have with each other when confined in animal and human tissue together. By themselves many are not a risk. But with all the different human exposures to chemicals today, many materials occur together with sometimes more damaging effects than if they appeared by themselves in living tissue.

Because people and wildlife share a common environment, they carry the same mix of persistent man-made chemicals in their bodies. In addition, higher members of a typical food chain tend to concentrate the amount of toxic chemicals from their many food sources. In Lake Ontario, for example, the tissue of herring gulls may contain 25 million times the concentration of PCBs found in the lake's water because they eat small fish by the hundreds that also contain these chemicals.[88] Their feeding habits tend to concentrate the many minute amounts contained in their prey. It is, therefore, not surprising that humans seem to be suffering increasingly from the same health problems reported in laboratory animals and in wildlife exposed to one or more of the dozen POPs.

Additional evidence of these patterns in nature comes from the study of bottlenose dolphins that have stranded and died along the Texas coast.[89] Researchers found as many as 209 different kinds of PCBs in their bodies. The toxic equivalency levels found in 10 dolphins studied were 200 times those known to cause birth defects in rats and developmental defects in birds. These findings raise serious questions about whether humans living in environments close to the sea or eating fish are also accumulating high levels of potentially poisonous PCBs. Some research is now suggesting that we may be even more vulnerable to PCBs than dolphins.

Problems from these POPs can include immune dysfunction, neurological and behavioral abnormalities, and reproductive disorders. Although the pattern of evidence is highly suggestive, it is virtually impossible to answer questions about the impact of these persistent chemicals on human health directly or definitively. In addition, because everyone carries a load of these chemicals, there is no unexposed population to study as an unaffected comparison group. Moreover, scientists for ethical reasons do not conduct experiments on people. Nevertheless, the weight of the evidence indicates strongly that chronic exposure to POPs is a hazard to human health that more than justifies precautionary action to eliminate them.

UNANTICIPATED CONSEQUENCES

Waste disposal sites, pesticide applications, and even storage of gasoline is causing toxic chemicals to move into our drinking water, whether it be in underground aquifers that we drill into or from the rivers and lakes we draw drinking water from. Present strategies assume that through various treatments and filtering processes drinking water can be "cleaned" of toxic chemicals. But there are now thousands of chemicals being produced without toxicological testing on humans or animals while governmental regulators are only required to test for 80 substances in water.[90]

Consider the MTBE situation that is happening around the U.S. now. The oil companies did a study validating the fact that burning gas causes air pollution. Thus, it needed an additive to more completely burn away the hydrocarbons that contributed to smog. The U.S. Environmental Protection Agency and several state governments decided to pass laws that mandated the use of an additive, if a good one could be found.

So the oil companies developed the very additive that their own gasoline needed to become clean and efficient…methyl tertiary butyl ether (MTBE). And then, as it turned out, the oil companies found that their production of MTBE could be formulated from a by-product of gasoline refining that they had earlier been throwing away. And like magic, this waste could now bring a yearly income of billions of dollars.[91]

Unfortunately there is a glitch, the stuff causes cancer and respiratory degeneration.[92] MTBE has been shown to induce liver and kidney tumors in rodents. It makes water taste like turpentine, leaks out of gasoline holding tanks, and gets into our ground water, our drinking water. Think about the City of Santa Monica, CA that had to shut down five of its wells in January, 2000…that is half of its water supply…because MTBE contamination was found from local corner gas stations.[93] As of March 2000 as many as 9,000 community water wells in 31 states may be affected by contamination from MTBE.[94] The real scary part of all this is that the oil companies might have already discovered a way to formulate gasoline so that it burns more clearly without the addition of oxygenates like MTBE or even ethanol. The whole additive question might be a sham that has resulted in potential exposure of millions of people to MTBE.

And, consider this, human populations, especially in the U.S., have very little exposure to heavily regulated areas such as Superfund sites, factories, and local industries. Virtually all environmental regulation, however, focuses upon these areas. And yet exposure to everyday consumer products like cleaning solutions, air fresheners, and building materials poses a critical danger to virtually everyone on a daily basis. Every year the average family disposes of 21 pounds of household hazardous waste. For example, 51-67 million gallons of oil from drivers are improperly disposed of annually, six times the Exxon Valdez spill. This risk might even be greater than that faced by those who live in communities surrounded by toxic industries. The evidence suggests that each individual must take a more proactive role in protecting their health. Environmental regulations cannot be relied upon to determine the safety of the places where we spend most of our time. What products we buy, where we live and work, and what we choose to eat become the determining factors in maintaining our health.[95] When do we have enough evidence to do something and when do we begin to err on the side of caution?

Chapter 5

The Way of Nature

"What we thought was boundless has limits and
we're beginning to hit them."

Robert Shapiro, Chairman & CEO, Monsanto, 1997

As CEO of Monsanto, Shapiro's surprisingly enlightened comment aptly states that we are and have indeed hit limits and are continuing to hit limits in a paramount manner with regards to the bounty of Earth. Actually limits puts it mildly, we are colliding into brick walls. Sooner or later businesses and communities around the world will hit the wall.[96] If Mr. Shapiro, the head guy of one of the largest polluters on this planet, admits that we are up against limits, that pretty much says it all.

PATTERNS IN GENERAL

To fully understand how humans fit into the bigger picture of the environment, it is helpful to look at the dynamics that make up ecosystems and how we fit into these natural systems that are all around us—all the little kingdoms we cannot see and the inhabitants of these mini-worlds. Learning to recognize the many life-supporting systems that occur in nature tends to begin with **awareness** of the smallest parts. It continues through an attempt to **understand** the system as a whole. *Webster's Dictionary* defines environment simply as "the aggregate of surrounding things, conditions, and influences." As we are citizens of the Earth, as well as citizens of a particular city, town, or county, knowing our home place by not only the road names and local landmarks, by not only where the nearest McDonalds, video rental stores, or local markets are, and by not only its history and its geography...but also by its natural cycles, is extremely important.

For example, we instinctively follow nature in a general way by the little things we do as we get ready for the seasons to come and go. Such activities as putting window air conditioners in, to preparing the soil for our gardens, or breathing a sigh of relief that the tomatoes seem to no longer be dictating our days, to bringing in or putting out yard furniture, to putting up Christmas lights or taking them down. Just as we adjust for the seasons of nature, so should we adjust for the various life cycles therein.

Understanding how these kingdoms work within the natural world, without destroying them is just as vital as understanding the importance of their existence. Understanding how the upland reach of a stream relates to the whole water catchment of a watershed is something we should understand for the areas where we live. Knowing where the water flows, how much gets distributed, and where it gets distributed, is vital. Most people in a community do not see that first-order water catchments are the initial controllers of water quality for supplies of their domestic water, the water they drink from their kitchen faucets. Listening and learning from those around us to what is around us is essential for survival at so many levels. It is as basic as simply knowing what is in our water, which is always questionable in this day and age.

As we become more aware of all the life nature holds around us, we realize that nature and the little mini-kingdoms or ecosystems that lay within are exactly what sustain and maintain us. And as we learn more about these little kingdoms that exist within nature's world, an awareness that some parts are very small materializes; and these small parts are a good place to begin exploring our relationship with nature. Just like building anything, the first few blocks are essential. A solid foundation for any system is vital. If the foundation is flawed or damaged there is a time-bomb ticking quietly away.

HOW IT ALL FITS TOGETHER

There are patterns of space around each of us. There are animals that move from place to place. There are plants that are sorted between different soils, altitudes, or water domains. And, there is a purpose for all those little creatures we are so quick to annihilate with Raid. The inhabitants in our surroundings change, just as we do under different types of conditions. All the living things, from the types of trees, to the birds that live in these trees, to the bugs that these birds eat, to bugs that the bugs eat, and so on. All change and have cycles that hold a purpose. And they all aid in sustaining the foundation of the nature.

The foundations of the ecosystems around us can be protected and shown respect by considering the following questions: What are the parts? How do they interact? What happens when some of them are missing? Are some of them ailing right now? What natural processes are or should be at work?

Children do this so naturally: always looking in the grass for odd things; keeping bits and pieces found outside in their rooms; looking a little closer at the things around them; looking a little closer at life; taking the time to live rather than just exist. Connecting with nature's ecosystems tends to breathe life back into the mundane. Interestingly enough, what we might all focus on to save our planet (those moment to moment activities) is exactly what might also save our sanity in this fast paced, high-pressure world.

Consider the story of declining underwater kelp forests in the Alaskan coastal Pacific Ocean that one way or another feed a range of species from barnacles to bald eagles.[97] The disappearance of these massive kelp beds caused governments and conservationists initially to hypothesize that pollution and other man-made disturbances were potential culprits. It turned out to not be that simple. In recent years food of pacific sea lions and seals has decreased, causing their population decline. These are a preferred prey of killer whales. With decreases in food, whales started preying on sea otters that live in the giant kelp forests along the pacific coast. The sea otters prey on sea urchins, which in turn are a major feeder on kelp. By the whales switching to sea otters for food, the otter populations consequently decreased, and their feeding was no longer able to keep the urchin populations in check. Thus, the kelp have been overgrazed by the urchins to the degree that these massive underwater forests are now disappearing.

Nature is quite helpful in letting us know what works and what doesn't. When something doesn't fit in nature it stands out like a sore thumb, such as the oil rigs in the middle of the Amazon forest. Nature is screaming this isn't a good idea, yet we proceed. We go after this "black gold" buried deep in the ground, through thick brush, in relentlessly remote locations, never stopping to question that perhaps it is buried so deep for a reason, that just perhaps it is meant to stay there. Then we become dependent on this "black gold" even though burning it sends untold amounts of pollutants into the air. Funny how those coincidences are always at work. Our lives would be much simpler if we listened to the hints nature sends.

Another example is our nation's wetlands. Our wetlands are vital to clean water, the non-erosion of shoreline soils, healthy and edible fish, and homes for crabs, migratory birds, and even otters. Many wetlands and their marshy areas are now being protected because of their vital role. The wetlands, marsh, and the life therein is one foundation amongst many that sustain us. Some people like to burn the marsh in an attempt to reduce the bug population, yet burning that marsh is not only a horrific offense on nature, but an attack on our own longevity as humans. That marsh is a mini-kingdom, and it supports other kingdoms both large and small. It is one of the foundations of nature and our world. And to destroy it for a better view or a longer dock makes that particular kingdom one Lego shy of a very vital castle.

There are many examples, just different foundations. The world's hardwood forests and rain forests are another example. An example that has yet to be protected from the negligent attacks of man. And a foundation that needs our attention, should we want to maintain any quality of life whatsoever.

As we've seen, the natural world around us not only works well together, but natures' components do not operate without each other. Everything works together for the greater good. Flowers need bees for pollination. The hermit crab needs marine snail shells for a mobile home. Several species of fish, clams, worms, and crabs live in the burrows of large sea worms, often eating the excess food or waste products of the host, keeping the worms tube clean. Lichens that grow on bare rocks are actually a mixture of fungus and algae. The fungus can not produce food and receives it from the algae. In turn, the algae can't hold water and receives this commodity from the fungus. Bacteria live in the stomachs of cows to break-down the cellulose of plants that cows eat for their nutrition.

Millions of bacteria also live in our own digestive tracks. They derive food from what we eat and help us to digest our own food. You often see white long-legged birds sitting on the backs of cows and pigs in farm fields. These white cattle egrets are benefitting from the insects stirred-up through the cow and pig grazing off the field grasses. Black birds eat cherries to obtain food. They also carry the cherry seeds considerable distances before excreting them, which in turn provides the opportunity for a new tree to grow far from its parent tree. The examples are endless and continually show the intricate relationships and coincidences that make up the strands in nature's web of life.

There is no doubt that understanding all that goes on with nature, in nature, and beyond nature can be extremely confusing. But if we can keep it simple and see nature as a life form and give it the respect that we ourselves would want as a life form, we understand at a level that comes from our hearts and souls. Likewise, relationships in the kingdoms of nature, just as with humans, can be complicated. Relationships are forever, evolving, changing, and effecting all that surrounds them and they in turn then begin their own changes. All are brought into order through unpredictable self-organization.[98]

Life of all kind is a continual, creative process. There are back-up plans that kick-in when change falls on hard times and there are supports for all levels of change. The kingdoms of nature, just as those of humans, not only are interconnected, and have interactive functions, but they can also be just as moody and full of coincidences as we humans. That tornado that misses a certain farm but annihilates a trailer park. That hurricane that is headed directly for a highly populated shoreline only to quietly turn back out to sea before massive flooding begins. That forest fire that takes pity on a neighborhood after destroying thousands of acres of natural parks. Nature's kingdoms are living, breathing, complicated life forces.

Learning how to live in "ecological realities," looking out for those valuable yet vulnerable smaller kingdoms of one's lifescape, and trying to duplicate what we can discover from their successful functioning, are the keys in our journey to sustainable development. Community members are citizens of a place that is living and dynamic, defined not by political boundaries or outdated legal codes, but by the ebb and flow of the natural world, its foundations, framework, relationships and its human connections.

NATURE IS NOT LAWLESS

The Earth operates with certain laws and rules. These core laws are in regard to the things of nature that we most rely upon: living resources and energy. The law of conservation of mass (first principle of thermodynamics) for example, simply states that mass can neither be created nor destroyed. Therefore, materials cannot really be "produced" or "consumed," nothing disappears. In other words, what is here is staying.

But, what many find confusing is that the form of what is here can be changed. The mass of a material remains the same while its form can be altered from a raw material to various finished products, wastes, and residuals without a change in how much is present. Think about how the wood from a forest is changed into furniture, but the basic makeup (*e.g.*, molecular structure) of the wood stays the same. Even the physical wood shavings from making the furniture are still present, only dispersed away as waste or compost from the original timber.

This first law, the conservation of mass, helps to explain how a collection of plants and animals, an ecosystem, carries on life and development. The ecology of an ecosystem is simply independent relationships of plants and animals competing and yet also knitting together co-development actions. Development is essentially the differentiation of a more general situation. Any given differentiation from a generality then becomes a new generality from which further differentiations can occur, causing the diversity of the system to expand considerably. But as this ecosystem develops the material and resources that initially allowed the animals and plants to inhabit the area and develop this ecosystem have not increased, just changed form and been more effectively used or reused by the many co-developing life forms of the system that have linked together in a large web of co-existence.

The law of conservation of energy is the second principle of thermodynamics. This law states that energy can neither be created nor destroyed, outside of a thermonuclear reaction. Thus, while energy can be changed in form and distribution, the amount remains the same. Energy is neither produced nor consumed; it is only converted from one form to another. And in this conversion, the disorder in the system increases so everything tends to disperse.

5 *Steel production plant on the shores of Lake Ontario, Canada, emitting large amounts of smoke into the atmosphere that contain both carbon dioxide and sulfur dioxide, contributing to the green house gases above the Earth.*

Think about it for a minute. If we burn a log of wood in the fireplace, the energy in this wood is transformed to heat. And where does this heat go? It does not stay in the log, but rather is dispersed into the room and up the chimney. Likewise, when the energy stored in fossil fuels (*e.g.*, coal) is released or "consumed", it is creating thermal, mechanical, or electrical energy, and bi-products, gases such as SO_2 and CO_2, which are pollutants dispersing into the atmosphere. Only the form of the energy has changed.

In more closely examining the complexity of ecosystem function and beginning to tease apart the many threads of its tapestry, it becomes easier to understand how the conservation of energy works so perfectly in nature. An ecosystem is like a conduit through which energy passes.[99] The system is initially fueled by the energy from the sun and then this captured energy passes through the conduit with few or many transformations, depending upon the plant and animal diversity of the system.

In an ecosystem such as a desert with few life forms, for example, not much happens to this captured sunlight energy. It hits the desert sand surface and most bounces off, back into the Earth's atmosphere, moving through the ecosystem's conduit very quickly. In contrast, the complexity, diversity, and numerous plant and animal co-developments of a forest ecosystem receiving the same energy flow from the sun cause this energy to move much slower and more effectively through the system's conduit. Thus, the diverse and redundant way that the forest system's many plants and animals use and reuse the energy, passing it around from organism to organism, means a very complex assemblage is able to be supported.

Through organism development and co-developments, nature has devised a means where energy is not destroyed (lost) but rather transformed and reused to run well-tuned "natural machines." The more different methods an ecosystem has for recapturing, using, and passing around energy before it leaves the ecosystem's conduit, the more valuable that piece of nature becomes.

NATURAL CAPITAL

The environment, all its little kingdoms or ecosystems, the nature in which all we creatures of the Earth exist, are comprised of what scientists call "natural capital." Natural capital refers to any stock or inventory of natural materials or resources found in our environment that yields a flow of valuable goods and services into the future.[100] This would be the forests of the world, a fish stock, or an underground water aquifer that can provide a harvest or flow.

Natural capital also provides waste receiving services, erosion and flood control, and protection from ultraviolet radiation. Thus, the ozone layer in our upper atmosphere is a form of natural capital, protecting us from harmful ultraviolet sunlight. Even thought the ozone is an 11,000,000 mile wide layer in the upper atmosphere, through our polluting actions on Earth we have increased the damage we experience from ultraviolet rays escaping through the deteriorating areas of the ozone layer. A crack in a natural foundation with clear consequences...skin cancer and global warming.

For example, when it is winter in the northern hemisphere and summer in the southern hemisphere, many cold-weary tourists from the U.S., Canada, and Europe flock to New Zealand for its many fabulous environments and radiant sunshine. But they may be getting more than they bargained for. In New Zealand over the last 10 years peak levels of skin-frying and DNA-damaging ultraviolet (UV) rays have been increasing, just as stratospheric ozone has decreased.[101]

Natural capital is broken down into two kinds. Stocks of oil, coal, or rich deposits of copper are some types in one category. We use these stocks of natural capital as fast as we want, but when they are gone, used up, depleted, and cast into the environment as wastes (better known as pollution), that is it. They will no longer be available for our use, for our children's use, or any generation there after. The inability to replace these fossil fuels is why the second law of thermodynamics is so important. It guarantees for example, that we can never take highly-dispersed atoms of say copper, and gather them back into a highly-concentrated copper deposit. The energy requirements of such an operation are simply too great.

Think about the thousands of years it takes to make a diamond from Earth's raw materials. If the second law didn't hold true, as Herman Daly from the University of Maryland says, "we could make windmills out of beach sand and use them to power machines to extract gold from seawater."[102] Unfortunately, the second law does hold true, and once we disperse highly-concentrated ores, we cannot afford to reconcentrate them. Thus, the mining of new materials from the Earth's crust must be seriously questioned. It should be diminished to the point that mining is hardly noticeable as a human activity any longer. Instead these materials from our inner-planet strata should be recycled as efficiently as we presently recycle gold. There are no waste buildups of gold anywhere in the biosphere because we recycle gold most efficiently due to its obvious monetary value.

The second kind of natural capital takes the form of a flow. In general these flows are continuous, though human bungling can interrupt some of them. Examples include sunlight, the capacity of green plants to create carbohydrates by photosynthesis, tree growth in a forest, rainfall, and the production of fish in the oceans. These forms of natural capital are endlessly renewable but can only be used at a certain rate; the rate at which nature provides them, either through natural biological or physical cycles.

For example, so long as we cut trees at a certain rate, and no faster, then nature will produce new trees fast enough to maintain a constant supply of cuttable trees. If we cut trees faster than that, nature will not be able to keep up with us and then people in the future will have fewer trees to meet their needs. The capacity of the Earth to support life will have been diminished. This is an example of exceeding the capacity of an ecosystem to regenerate itself; thus taking a perfectly solid foundation of nature and reeking havoc on it as well as havoc on ourselves.

This natural capital includes not only all the natural resources and waste sinks needed to support our economic activity, but also the basic physical life-supporting "services." These services include such things as the air we breath, the water we drink, and the food we eat. Natural systems need to be somewhat intact to do their job. Thus, all the little kingdoms need to be sound to produce our life sustaining needs. If there are cracks in the foundation that cause certain parts to collapse, we can expect problems. Society's health and prosperity depends on the enduring capacity of nature to renew itself and rebuild waste into resources. If the structure fails due to imbalance, nature and its ecosystems will fail to meet our needs of air, water, and food effectively. And these needs being affected will certainly initiate a flood of complaint letters that perhaps only Mr. Shapiro of Monsanto could relate to.

WASTE TO FOOD

Not only does nature provide high-quality raw materials that we use, but nature also biodegrades our biodegradable wastes so they can again become raw materials. But, this does not include plastic and all its derivatives.

For example, when we throw away wood, natural agents called "decomposers," such as termites, begin to eat the wood waste and break it down into raw materials—carbon, hydrogen, oxygen, nitrogen, sulfur, and so forth. Creatures such as earth worms use the termites' wastes as raw materials for food and to make soil, which provides nutrients for new trees to grow. This is called the "detritus food chain" and it is essential to life on Earth, though largely invisible from human view. It also maintains the first law of thermodynamics, talked about above...natural materials don't disappear; they just change form. Therefore, all the basic elements of that wood we throw away are just reformulated into other forms. But this process takes time and can be disrupted if more elements are consumed than reformed.

The detritus food chain is made up of insects, bacteria, funguses, and other creatures that most of us know little about. But without their workings, the world would become overloaded with wastes and biological processes would become clogged and stop working. Therefore, this detritus food chain is a very important foundation for life as we know it. If you've ever visited a modern hog farm, you have an idea of what it means to exceed the capacity of the local environment to absorb waste. It falls most definitely on the unpleasant and hazardous end of the spectrum.

In nature every last particle contributes in some way to the health of a thriving ecosystem. Waste from one form of life equals food for another life form. Take the cherry tree as one example of nature's industry, which operates according to cycles of nutrients and metabolisms.[103] It makes thousands of blossoms just so that another tree might germinate, take root, and grow. Who would notice piles of cherry blossoms littering the ground in the spring and think, "How inefficient and wasteful?" What blossoms don't germinate, however, simply decompose and produce nutrients for more tree growth. The tree's abundance is useful and safe.

The hog farm, on the other hand, is an example of a system that operates outside the laws of nature. The cyclical system of those that do operate within the laws of the natural world are powered by the sun and constantly adapt to local circumstances. Waste that stays waste simply does not exist. And, there is not an ounce of hazardous waste in this natural system. Once again nature is sending a huge hint as to how to live, recycle, nurture, and procreate.

NATURE'S APPROACH TO INSURANCE

The natural capital of every ecosystem on Earth also constantly faces the impacts of change, sometimes being over-abundant and other times stressed. The degree of a system's adaptability depends on the richness of its biodiversity, or the number of different plants and animals found in one place. Biodiversity creates a redundancy—duplication or repetition of the elements of a system. Thus, in case of a failure in one part, the system can still work, and retains the ability to respond to continual change.

How does this happen? In the framework of nature's redundancy more than one species can perform similar functions that will keep an ecosystem running—an ecological insurance policy. And it is these eco-insurance policies that provide flexibility to absorb change or to bounce back after disturbance. It is this back up plan that ultimately strengthens nature's ability to retain the integrity of its basic relationships. The insurance of redundancy means that the loss of a species or two is not likely to cause the collapse of all the little kingdoms therein, because other species can make up for the functional loss. The castles have back-up generators, and emergency plans in place.

Natural processes, however, place limits on what plants, animals, and even humans can and can not do. The most effective advantages to be gained from these processes are by those systems where life forms develop and co-develop through differentiation and combinations, where there are multiple uses and reuses of energy, and where life is maintained through refueling from within the system. In well functioning natural ecosystems these activities are interlocked and the entire process is referred to by ecologists as dynamic stability.[104]

But there does come a point, a threshold, when the loss of one or two important species may in fact tip the balance and cause the collapse of kingdoms, and in turn cause the ecosystem to begin an irreversible change. That change may signal a decline in quality or productivity, and ultimately less available for other species, including humans, that may rely on this ecosystem. A good example of this is the excessive amounts of wastes from factory farms that run into surface water systems. The water is then flooded with massive amounts of nutrients which in turn increase the algae growth and decrease the oxygen in the water resulting in massive fish kills. At the same time there is a new kingdom created that supports a very lethal bacteria. A bacteria that causes sores on the skin and confusion in the mind. Not a kingdom one would want to visit.

Another, more human oriented example, could be the newly elected mayor of a city whose budget is overspent but guarantees to balance it.[105] All that is necessary is to eliminate some services whose total costs add up to the over-expenditure. So the mayor gets rid of garbage collection because without this cost, he not only meets budget but has a bit left over. But see the problem that arises? Removing one part from the whole puts very smelly cracks in the foundation, causing chaos and probably eventually loses the mayor his job. Watching that garbage magically disappear weekly runs very close to most American's hearts. Point being, a simplistic view of looking only at the cost of and not at the function performed by the service is a huge mistake. The diversity of city services is essential for the city to maintain the life that resides there. Sound similar to the calls for maintaining nature's biodiversity?

Nature has a "plan A," a "plan B," and sometimes even a "plan C." It is these plans that strengthen the ability of all those little kingdoms to hold their foundations together in the context of an ecosystem. To remove a piece of the system may be acceptable, provided we know which piece is being removed, what it does, and what effect the loss of its function will have on the stability of the system as a whole. To blatantly toss foreign, generally toxic chemicals into the Earth, kill off large parts of it randomly, flood the Earths' waters with massive amounts of animal wastes, and continually show little respect for nature's life force, however, is like wiring all the kingdoms with TNT and leaving a chain smoker to mind the various detonators.

Why nature's "plans" and the many coincidences that present themselves work in the natural environment are because of the fact that no dynamically stable system lasts forever.[106] Even without the influence posed upon most natural systems by humans, dynamic ecosystems are always in danger of succumbing to instability and collapse, which is why they need to exert internal self-corrections. Of course luck plays a part, depending upon how plants and animals respond to the many coincidences that confront them. But dynamic ecosystems must continually correct themselves using the resources and methods they have available to them in their environment. These corrections fall into the categories of bifurcations (forks in the road) and feedback loops.

A system's instabilities could become so serious that for it to not collapse would require a radical change, taking a new fork in the road of its development. This is what evolution of species is all about, such as animals originally living in the water (*e.g.*, the lung fish) and then developing a primitive air breathing lung that would allow them to move onto dry land, maybe to avoid a predator that could eliminate their species. Likewise, in the evolution of human cultures, Romans for example built aqueducts when local springs and wells dried up.

Innovative hunting cultures began saving captured game for breeding, a correction to uncertainties and instabilities of the hunting life.

Feedback controls in ecosystems restore balance among the parts. Feedback messages race inside plants and animals, for example so predators don't destroy their prey sources. In addition, feedback messages are at work in the insect world. In termite colonies all eggs produced by the queen are originally the same. But they develop into the different adult types (*e.g.*, soldiers, nursers, etc.) based upon densities of existing adult types and the chemical odors they give off, which signal the eggs as to which adult types are low in number in the colony and need replacing.[107] Human breathing also illustrates the example of feedback loops. One knows when to take another breath because of the rise in carbon dioxide level in the bloodstream. This in turn triggers the brain's breathing center to shoot a message to the diaphragm to contract and allow our lungs to fill with another breath.

UNDERSTANDING OURSELVES, OUR EARTH

Our connection to the Earth involves a very spiritual, innate, and resilient sense. This is why people in cities tend to visit the parks and zoos more than any other locations. This is why a walk in the woods can alleviate depression. This is why a child can pick up a small rock along a path, sit down on the ground and play with, investigate, and appreciate this rock for an indefinite period of time.

The value of nature lives somewhere in all our souls. The affinity towards the Earth we live on lies in emotions and feelings that people have for an area as much as the food or resource we get from the Earth. Our connection to nature can be a very personal thing, and in fact is derived from our spiritual awareness of our environment. Our honor for nature cannot be recognized in the narrowness of science, for honor is born in our hearts.

6 One day you pass this forested location and there is a wonderful wooded hillside habitat. A month later this is what is left after a logging company has clear cut the area for wood.

As nature dies our hearts sadden. And it is in this sadness that there is a growing anxiety we are feeling as our innate awareness awakens to all that we are losing in the gradual destruction of the smaller worlds around us. These changes, the kingdoms that are lost daily, are happening at a cut-throat pace and leave in their tracks a growing sense of regret. The changes run deep and attack the many different facets nature holds. Continuity continually destroyed. That woods we have passed for 10 years, then one day loggers are there, tearing out its heart, making many who live there homeless. As larger numbers of the kingdoms are destroyed, grief hits us square in the face. As the wildlife scatters and the forest stands raped, an empty essence now sits where beauty and life once existed. Another part of our life is now made barren. Depression rates continue to soar.

This depression and anxiety running through us all is a common thread, a cosmic hint letting us know that we are allowing things to be lost which should be protected for today and forever. But to simply medicate and rationalize fails to stop the sabotage of the Earth, both on a small and large scale. Ultimately we fail ourselves, our children, and our ancestors when we ignore the everyday changes needed that would make lifestyles more sensitive to the needs of nature.

Yet, in survey after survey there is a public expression of very strong feelings for nature. It ranks second only to freedom. This expression comes from deep inside, that small whispering inner voice saying to each of us: "as you kill the Earth, you slowly, but surely are killing yourself." We know at some level that our castle walls are beginning to crumble.

Environment must move from our sub-conscience to consciousness, integrating into all areas of life. The environment, however, remains largely outside the mainstream of everyday thought, still being considered a sort of add-on to the future of life.[108] If impotent in understanding what we as members of the largest kingdom of nature might do to save our castle, self-destruction is inevitable. Lurking around each corner of hope there continually appears that uncrossable gulf between the things we care about and those things which are recognized as essential to life by the politicians, decision makers, and destroyers of our life supporting kingdom.

As we tread through a national administration destined to reduce aide to the environment, drill and destroy the last Arctic Refuge, ignore Global Warming, open up protected lands to more destruction for corporate financial gain, and dismantle and deceive global alliances set in place to protect the world from the fallout of excess, it is without question a time to take back our land. As people of consciousness, the honoring of nature will not be done by war, or violence, or by further destruction, bloodshed or slavery. We shall take our land back by knowledge. We will learn what it means to live harmlessly with the natural life around us and we will spread the word. One by one we are realizing our power. We are not of weak mind, or spirit. We are reconnecting to our core foundation, the Earth, and learning about these very fragile castle walls. It is indeed time for us, one by one, to honor our kingdom, defend our castle, and protect all the smaller subjects therein.

Section III

Why are People and the Earth at Risk?

Chapter 6

Outside the Web of Life

"Beauty is the adjustment of all parts proportionately so that one cannot add or subtract or change without impairing the harmony of the whole."

Leon Battista Alberti

In 1776, the world was essentially empty from a human perspective, with fewer than one billion human inhabitants. At that time, the planet had abundant "natural capital" of all kinds—for example, highly-concentrated metallic ores, oceans full of fish, continents covered with trees to absorb carbon dioxide from the atmosphere, and mysterious substances like petroleum oozing out of the ground spontaneously. The world of 1776 was short of "human capital"—techniques for extracting minerals from the deep earth, ships to catch fish efficiently, and machines to turn trees into lumber. Now, says economist Herman Daly, the situation is reversed.[109]

OUTGROWING EARTH'S CAPACITY

Rapidly progressing, major changes such as global warming, ozone depletion, soil degradation, forest depletion, and species extinction threaten the delicate ecological balance of the entire globe, emphasizing nature's constraints on human activities. With the closing of the twentieth century, the threat of irreversible degradation of planetary life systems by these and other possible unanticipated dangers has come to replace nuclear war as the primary concern of society.

But, despite technological, economic, and cultural accomplishments, we remain first ecological beings and a part of the larger natural world. As ecological beings, and especially very intelligent beings, we affect five of the seven foundation elements of survival: air, soil, water, biodiversity, and population density. We affect the other two, sunlight and climate, indirectly.

Thus, to achieve sustainability, people in a community begin to respect the land on which they live. A basic land ethic running parallel with a basic human ethic. Then they develop respect for natural laws, acknowledging without question, the first law of ecology, that everything is related to everything else. And, everything must exist in relationship to something else. Conversely, nothing can exist out of relationship to something else. The unified world view is predicated on the notion of holism, in which reality consists of organic and unified wholes that are greater than the simple sum of their parts.

Forests, soils, wetlands, lakes, oceans and other naturally productive ecosystems provide food, lumber, habitat, oxygen, waste handling, temperature moderation, and a host of other essential goods and services that are considered natural capital essential to human existence. For millions of years they have been purifying the planet and creating a place suitable for human and other life.

Sustainable development emphasizes that people live within the capacity of the renewal ability of these materials, goods, and services from the natural world, our environment.[110] That means only taking the amount of natural capital that will not affect the capital itself or, as some would like to say, living off the interest of our natural capital like a bank account or dividends from our stock investments. When we degrade these systems by using more natural capital than Earth's capacity to supply, we endanger our livelihoods, undermining the worth of our bank accounts and the likelihood of our continued prosperity and existence.

But this goes even deeper than one would first imagine. It involves what environmental science recognizes as a "resource buffering theory."[111] For every resource essential to a life form where a very small proportion is directly used for life processes of individuals, the vast remaining proportion of the resource is indispensable to maintain conditions under which the population as a whole can continue to survive. For instance, although we need only about 1.5 to 2.0 quarts of water per person per day to stay alive, the total human population needs the balance of the water resources in the atmosphere, oceans, ice, wetlands, and other aquatic systems to buffer for emergencies. These masses of water provide crucial functions by absorbing and redistributing energy and waste products from life forms. They shield us against the atmosphere's fluctuations in gaseous content. And they offer transportation and provision of conversion sites for nutrients. If such resources are spoiled, conditions for human life will inevitably deteriorate…a person's 1.5 to 2.0 quarts of water won't save them because we don't know specific quantities that constitute a sufficient buffer. Policies must reflect and be built on the conservative natural distribution and use of the world's total resources.

Increasingly natural capital is becoming more scarce and human capital is expanding or humanity is outgrowing the Earth. Natural capital is a pre-requisite for human-made goods. The ingenuity of human capital cannot substitute effectively over the long-term for natural capital. What good is a sawmill without a forest? What good is an oil refinery without oil? To put fish on our dinner plates, both a fish stock and fishing boats are needed. All the fishing equipment and processing plants in the world (human capital) will not generate a single fish if the natural stock is gone. Consider the following.

1)   Today there is no shortage of huge ships to sweep nets through the oceans to harvest fish—but the fish themselves are disappearing.

2)   Chemical factories are abundant, producing a cornu-copia of useful chlorinated chemicals, but there is a shortage of natural mechanisms to detoxify and recycle such chemicals. As a result, the entire planet is experiencing a buildup of toxins, such as pesticides, and scientists are discovering new harmful effects in wildlife and humans each year.

3)   Only recently, scientists concluded that the ecosystem's capacity to remove carbon dioxide from the atmosphere has been exceeded because of human activity. As a result, they believe, CO_2 is building up in the air, pushing up the temperature of the planet, which some have calculated has increased as much as 1.4° F in the last 100 years.[112] We are waiting now to learn the real consequences, but more droughts, floods, and major storms must be expected, we are told.

Let's take a closer look at the idea of human capital, in contrast to natural capital. In the age of the Internet, consider for example communication wires. Wiring used to be made of copper ore. But copper is in short supply. So sand and large amounts of energy have been combined with massive know-how to manufacture glass fiber filaments. People see this as an example of our ability in using technology to overcome problems, the development and growth of human capital.

But what did it cost our resources to advance this technology? Where did the sand and energy come from? It takes a lot of energy to turn sand into glass. And, you can bet it wasn't green energy. Most likely, the energy source was generated by burning gas, coal, or oil, which in turn threw untold amounts of toxins into the air and helped global warming along. In other words, fiber optics, a relatively harmless product in themselves, probably aren't as good for the environment, in reference to resource use and pollution reduction.

THERE IS NO SUBSTITUTION FOR NATURE

Mainstream economists do not worry about shortages of natural resources to supply our needs and receive our wastes because classical theory assumes that human resources can substitute for natural resources.[113] To a limited extent, this is true. We have always been conditioned through history to realize that human tinkering can always come up with a solution to limited resources. Woven cloth and rugs were substitutes for animal skins. Columns of quarried stone and kiln-made bricks were substitutes for timber.[114] If we ever stop tinkering to correct the mishaps happening in our world from disappearing natural resources, we might begin to realize the real affect of these lost resources.

Herman Daly, however, argues traditional economists ignore the extent to which the usefulness of human capital depends upon the availability of natural capital.[115] Probably the most revealing fact of human's relationship with the environment is to consider the issue of "throughput beyond environmental regenerative and absorptive capacity."[116] Throughput is the flow of materials and energy through the human economy. It includes everything we make and do. When we speak of "growth" we are talking about growth in throughput—people making (and throwing away) more stuff and using more energy to do it.

The totality of the human economy is measured by throughput. It is calculated as the total number of people multiplied by their consumption and waste. Thus, there is consistently a dependence of economic activity on human and natural resources, in addition to physical and financial capital. Economic activity must not irreparably degrade or destroy these natural and human resources. Some material goals are inescapable, but in a world of limited resources, unlimited material goals clearly pose a problem.[117]

There is considerable evidence now, however, that the use of natural capital by many parts of our economy, in the process of throughput, has already exceeded the regenerative and absorptive capacity of the environment.[118] The problem of climate change and global warming are commonly reported examples. These issues provide evidence that we have exceeded the capacity of the atmosphere to absorb our carbon dioxide, methane, and nitrogen oxide wastes.

There is also a shortage of natural mechanisms to detoxify and recycle harmful chemicals from the environment. Many of the fresh water fish of the world now contain dangerously elevated levels of toxic mercury because we humans have doubled the amount of mercury normally present in the atmosphere. Have you ever seen what deformities mercury poisoning causes a human being? They are extensive and dehabilitating, one of the saddest man-made tragedies, and one that is as sporadic as it is dehabilitating. This poison is major evidence that we have exceeded Earth's capacity to absorb our mercury wastes, which remember are natural to the Earth in small quantities. Likewise, depletion of the ozone layer is evidence that we have exceeded the atmosphere's capacity to absorb our chlorinated fluorocarbon (CFC) wastes.

Garbage as our most basic waste also has so many dimensions, whether it is garbage dumped in the darkness of rural America or garbage placed neatly along polished suburban streets. Annually Americans discard an average of 1,500 pounds of garbage. And, once garbage is out of our sight does not mean that it is out of our lives. There is some garbage that evolves into harmful products re-used by nature and some that doesn't. Sadly, more than 50 percent of garbage placed in landfills could have been recycled by humans.

Once the garbage has left our sight nature begins to take it apart and recycle it, if it can. Hazardous, which is one type of garbage that nature is helpless against, is a type the doesn't go away. Hazardous waste can be everything from chemicals that once were used on soil and finally have been dubbed illegal and stored negligently in barrels that rot, to nuclear waste which has that radioactive edge to it, to farm factory waste which is the excrement of millions of abused animals and has more strains of bacteria and other little nasties than one could come to grips with.

Some non-hazardous waste doesn't go away either, such as Styrofoam. And, we are creating so much waste in the U.S. that soon the waste site outside New York City will be so high it will need an aeronautics license to notify pilots. What part of this problem do we not understand? How can we plead "I didn't know" to this little monument of our wastefulness. Aside from the blind, the evidence most definitely presents itself.

We have severely imposed on the basic cycles of nature that provide the foundational elements of our survival by not respecting the interactions and activities of all the smaller kingdoms there in. We have greatly affected the regenerative and absorptive capacity of the environment, which refers to the ability of the environment to provide materials for our use, and places where we can throw our wastes. And we are robing ourselves of nature's greatest gift, the ability of the environment, and the kingdoms there-in, to break down our wastes and turn them back into raw materials: an astonishing piece of work and a noble, vital service.

IS CLIMATE REALLY CHANGING?

There are way too many humans and way too few trees. Clean air is scarce, and pure, untreated water is even scarcer. We are outgrowing the Earth's supply of natural resources and waste receptacles. For example, most of the world's seventeen marine fisheries are badly depleted—a flow of natural capital that we have over-harvested, in some cases nearly to the point of extinction.[119]

And what about this thing called Global Climate Change due to warming of the Earth's atmosphere? The climate is forever changing and has always been a topic for park-bench discussion. The climate does its thing and we adapt. Its not like humans can actually control the climate, right? Well, human activity has caused the Earth's temperature to rise by more than a degree in the last century. And this change has the world up in arms.

The French physicist, Jean Fourier, argued 170 years ago that the Earth's atmosphere acts like the glass of a greenhouse, labeling this the greenhouse effect.[120] The greenhouse effect is a natural phenomena that keeps the Earth about 60 degrees warmer than without it. Without it the temperature on Earth would be around 0 degrees Fahrenheit. After the sun's radiant energy passes through the Earth's atmosphere, like it would a greenhouse glass roof, it warms the ground of Earth (inside the greenhouse). The heat then reflects off its surface and is trapped below the atmosphere gases (under the greenhouse roof glass), preventing it from escaping back into space, to create a nice and warm place where abundant plant life can flourish, just like in a greenhouse.

The problem arises when human activities, like driving our cars, burning coal to heat our homes and run our factories, chopping down forests to build our cities and produce paper, and raising cattle to fill our bellies have significantly increased the atmospheric concentrations of key greenhouse gases like carbon dioxide, methane, and nitrous oxide. Compare this to the idea that all of sudden our greenhouse glass roof design has changed. The properties of the glass have been changed so that they don't interfere with the passage of radiant sunlight through, into the greenhouse, but these properties are better insulators against escaping heat, keeping more heat inside the greenhouse.

Is this warming really occurring? According to a recent study of temperature readings interpreted from geologic core holes drilled into the Earth's surface, a 500 year warming trend for the Earth has accelerated in the latter half of the 20th century.[121] Over the longer time of the last thousand years the temperature of the Northern Hemisphere has been dropping slowly (approximately 0.04 degrees Fahrenheit per century). That is, until around 1900. Since then there has been a dramatic temperature rise, with 3 of the last 5 years being the warmest in instrumental history (see the graph).[122]

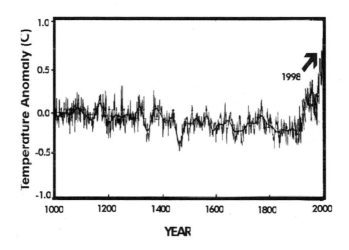

7 *The millenium cooling trend suddenly reverses (from The Mercury's Rising, 1(1), 1999). In the last 1,000 years the average annual temperature of the globe's Northern Hemisphere has been dropping slowly, that is until 1900. Since then there has been a dramatic temperature rise, with 1990 the warmest year known since 1400 AD. The plot above shows the average movement (anomoly) of annual temperature away from a 1,000 year mean (represented by 0.0 on the X-axis).*

Carbon dioxide (CO_2), the most significant human-caused greenhouse gas, is produced primarily by burning coal, oil, and natural gas. This burning releases about 5.5 billion tons of atmospheric carbon yearly. Another 1.5 billion tons is released through land-use changes such as deforestation. Plants and the ocean currents absorb about half of this carbon dioxide. The rest remains in the atmosphere for 100 years or more. CO_2 in the atmosphere has increased more than 30 percent from pre-industrial levels. To stabilize atmospheric carbon concentrations at no more than double the CO_2, many global climate change experts recommend a cut in emissions by 60 percent or more.[123]

Although it is not absolutely clear what all the factors are causing this phenomena of warming, there has been a discernable human influence on global climate.[124] The realities of this warming in terms of different effects on our world is becoming quite clear. For example, both the European Alps and Caucasus Mountains have lost half their ice in the past century. Global sea level has risen by at least four inches this century. Plant species are growing higher on mountain slopes. Growing seasons in Europe and North America have increased by approximately eleven days and heat waves, droughts, and intense precipitation events are more common and more severe.

Conclusions of a new federal report, "Climate Change Impacts on the United States," suggest a mixed bag of effects of global warming on the United States:[125]

- Average temperatures will probably rise 5 to 10 degrees Fahrenheit—nearly twice the projected warming for the planet as a whole—prompting more summer urban heat waves and gentler winters across the nation.
- Agricultural production will likely surge, and forests will probably flourish, thanks to the fertilizing effect of more carbon dioxide in the air. But many long-suffering ecosystems, such as alpine meadows, coral reefs, coastal wetlands and Alaskan permafrost, will likely deteriorate further.
- Snowpack will probably diminish by 50 percent on average, while winter rains are likely to increase, bringing 60 to 100 percent more showers to much of Southern California and the parched Southwest.
- Total precipitation nationwide, which rose 5 to 10 percent during the 20th century, will probably increase another 10 percent by 2100, chiefly in the form of extreme storms.
- The threat of drought—especially in the western Kansas and eastern Colorado breadbasket—will rise because hotter conditions enhance evaporation.
- A doubling or tripling of heat-related deaths in Minneapolis, Chicago, and other cities that rarely experience extreme high temperatures. The July heat index is likely to rise by 10 to 20 degrees in the mid-Atlantic region.

Continued global warming will affect far more than the natural environment. Public health, real estate, commerce, and global environmental policy will all feel the heat over the next 100 years...making climate change an issue that effects everyone.[126] Heat waves, which have already increased by 21 percent since 1949, are expected to increase the frequency of heat strokes. A greater demand for air conditioning, puts demands on power plants, which in turn will use more cooling water, putting more demand on scarce water resources...not to count additional emissions into the atmosphere from fuel burning to produce the power.

Global warming may also be triggering super storms. Eight of the 10 most destructive hurricanes since 1953 have struck the U.S. in the 1990s.[127] Melting ice caps, sea level rise, and more frequent storms...all consequences of projected temperature rise from the global greenhouse effect...could cause catastrophic floods.[128] Flooding can bring pollutants and waste into relatively clean areas. In addition, consequences for real estate are likely to be serious. With the projected 39 inch sea level rise in this century, 49 feet of shoreline may be lost. We will lose 450 feet of New Jersey shoreline this year from the melting of the polar ice-banks due to this warming.[129] Granted New Jersey is New Jersey and one of a more toxin ridden states, but still, beach is beach. Even more frightening, especially for the drug runners (finally), is the fact that by 2030 most low-lying areas of Florida, which is essentially most of the state, will be completely underwater. That old joke about buying swampland in Florida takes on a eery essence.

Western alpine forests also are likely to disappear, threatening the attractiveness and functional importance of natural environments, not to mention the timber industry. The place you hope to build your vacation home might someday not be there. With global warming, the oceans are also getting hotter causing:

(1)   salmon stocks to not find suitable habitat in the Pacific Ocean;

(2)   fish and other species to shift toward the poles in response to warming; and

(3)   the decline in sea ice at the poles, depriving birds, whales, and polar bears of their hunting and breeding grounds.[130]

In fact, the more scientists look, the more they observe relationships between shift in climate and changes in animal behavior and populations.[131]

Extreme weather events more numerous now, also have a profound impact on public health, especially in developing countries.[132] Health officials have been able to predict epidemics of Rift Valley fever up to five months in advance by tracking rising ocean temperature and increased rainfall. Other disease outbreaks related to more sever weather from global climate change include cholera, malaria, and dengue fever. Cholera epidemics can now be related to climate and climate change events, including ocean warming such as El Nino.[133] The cholera bacterium is a parasite of a common plankton (microscopic aquatic plants), and warmer waters will mean more hosts and more chances for cholera to proliferate.

Therefore, whether or not we can pinpoint the exact culprit or culprits causing global climate change, doesn't the evidence of effects behove us to start doing something in lieu of suspicions? Governments might just want to begin taking the lead and reducing the emissions of greenhouse gases released into the atmosphere when fossil fuels are burned and automobiles are driven. Otherwise, many of us who have things today will be the have-nots of tomorrow.

Ironically, fossil fuels may have done more damage to our people, our environment, and to our perceptions of our relationship with the natural world than any other material. Think about it: our ability to get into a car and go anywhere we want anytime we want has dramatically affected our very perception of fundamental concepts like time, distance, and space. We measure distance in terms of freeway minutes and consider our car an extension of ourselves. But how much land in U.S. cities is devoted to cars (i.e., streets and parking lots). Would you believe 33 percent.[134]

AWARENESS AND SUSPICION

In our interactions with others, we must seriously think about how activities impact not only our own backyard but the global backyard. Consider again the fact that we use the global atmosphere as a receptacle for local wastes. Chris Maser tells the story of the people of Canadian First Nation living sustainably within their ancestral lands, at the northern edge of the Canadian boreal forest.[135] Their livelihood was hunting, fishing, and trapping fur-bearing mammals, as they had done for centuries. Enter airborne sulfur dioxide from the cities of southern Canada and the northern United States, over a thousand miles away. The resulting acid deposition begins in secret to kill the boreal forest. Gradually, the forest deteriorates to the point that it alters the habitats of the animals on which the people have traditionally relied for food and trade. With the decline in forest health comes poverty and destabilization of the people's culture. With nowhere else to go, the culture collapses through no fault of the First Nation peoples.

Few people know much about the natural world around them. We often do not know what is grown on the land in the community in which we live. We are not aware of how it is grown, and how that effects us. And, amazingly enough, many are not the least bit suspicious when we see skull and crossbones signs in the fields around our homes. Suspicion seems to be a basic here, wanting to know why that sign is there seems important. Wanting to understand all the connections that exist with that field, that sign, that forest that was wiped out to create the field in the first place, is probably more interesting than any TV show that is on.

But, as we coat everything with black tar and plastic we become more and more removed from the little kingdoms and their dance that support our lives, and which are vital to life of all kind. We have lost an innate honor for life in general. You often hear from many when honor for other life is discussed: "why should we?" Repeated more times than anyone could ever remember, "We are after all King of the beasts." Back to that divide and conquer mentality that masks so many's sense of aloneness and emptiness and also tempers our need for asking questions. And yet, we know enough to be dangerous. We know if we spray "this" on "that" it will die. Is it time for our suspicions to become more instrumental toward influencing our thinking and actions?

Darwin long ago hypothesized that English cat lovers might unwittingly be setting off an ecological domino chain effect that led to prettier gardens. Cats eat mice that normally pillage the nests of bumblebees, so Darwin reasoned that more cats would mean more bees—and more of the red clover and purple-and-gold pansies that bees pollinate. Thus, the more cats, the prettier the gardens in a district.

And our suspicions can even take on a more profound life and death meaning. The day-flying moth, *Urania flugens*, found in Mexico and South America, metamorphoses from a caterpillar that feeds exclusively on a particular variety of trees and vines know as *Omphalea*.[136] The heavy defoliation caused by the feeding of the caterpillars in turn causes the plants to produce a protective chemical toxin, which makes them unpalatable to the moths. The plants' toxic compounds have also been found to be effective against the AIDS virus in test-tube experiments.

There is a problem, however. The toxin is produced only when a plant interacts with a large population of caterpillars. By simplifying the composition of plant and animal species, the timber industry in cutting down much of the forest for example, simplifies the structure of the forest, converting it into a tree farm, and minimizes the capacity of the moth to reproduce in large numbers. Such simplification removes many of the interactive, interconnected, interdependent functions on which the long-term stability and adaptability of both the forest and people depend, seriously disrupting natural, unpredictable self-organization.

A team of researchers from the Institute of Ecosystem Studies, in Millbrook, NY and Oregon State University in Corvallis, OR, studied connections among white-footed mice, ticks, gypsy moths, deer, and

Lyme disease.[137] They found that in upstate New York forests in years when there were an overabundance of acorns produced, there were also booms in the mice population because they eat acorns. Mice also eat the gypsy moth larvae found in nests in the trees. So when acorns produced from trees were abundant, the mice were abundant and kept the gypsy moth populations in check, eliminating their threat for plundering Eastern U.S. forests.

White-footed mice, however, also carry in their blood the Lyme disease spirochete, which they transmit to tick larvae they become infested with from the forest floor. When there is an overabundance of acorn production, tick-bearing deer are also attracted to these sites, for the acorns they like to eat, where the mice populations have boomed. The adult ticks, on the deer that gather in larger than usual numbers to eat the abundance of acorns, spawn more larval offspring which infest more mice, and thus more ticks pick up the Lyme disease vector. So while the plundering of the gypsy moth is being kept in check by one series of ecological mechanisms (mice feeding), the dreaded Lyme disease also has the potential to proliferate.

A community, humans, animals, plants, land, air, and the water that surrounds it, are joined in a common landscape. A human community's world view, its suspicions and resulting awareness, defines its collective values, which determine how it treats its surrounding landscape. In turn, the surrounding landscape and its environment is the community's social mirror.[138] The condition of the landscape and its environment reflect how people are acting.

AN EVOLVING SEPARATION

History repeats itself. Cycles and revolutions seem to continually show themselves in some form. Copernicus awoke us to the fallacy of human centeredness in the universe centuries ago. Though Galileo delivered more proof that the sun did not revolve around the Earth, it was years before the accepted paradigm integrated this less egocentric view. Humans could no longer presume themselves to be at the center of the universe. Copernicus and his followers thus challenged the very cosmology of Christianity.[139]

Truth sets minds free—throwing open doors for research and discovery, laying groundwork for great new structures of scientific inquiry, sparking a renaissance of action. The world would never be the same.

History has a funny way, however, of repeating itself. Cycles and revolutions seem to continually show themselves in some form. Copernicus was eventually able to put-down the long-ago believed philosophy of human centeredness in the universe. This egocentric status is now repeating itself—though not with the concrete features of the era in which Copernicus existed. Now the issue involves new latitude. Humanity's present day view involves being separate from nature. The self-centered perception that nature exists solely for the purpose of humans is comparable to the idea that the sun revolves around the Earth.

As human beings, "created in the image of God," we have come to believe we occupy a unique station in creation. We have evolved through all the physical kingdoms and contain all of their capacities, plus our distinguishing capacity for rational and self-reflective thought. This unique capacity of the human mind, allows us to mediate between the material and spiritual dimensions. This has caused us to separate ourselves from nature, both externally and internally.

This was probably first and best represented by the comings and goings of the great civilizations of recorded history (i.e., Greeks, Romans, etc.). In the evolution of these cultures there was increasing emphasis on the rational mode of consciousness. The drive was towards greater independence, order, and abstraction representing the primacy of human and masculine energies. Nature was gradually demythologized, earlier views of harmony were abandoned, and spiritual and intellectual pursuits were abstracted from the world of nature and its instinctive primal energies. Nature began to be subsumed as a resource for the development of larger collective units of social human organization. Trade, commerce, and artistic, as well as intellectual pursuits, were associated with urban dwelling, further increasing physical separation from nature.

We have continued this basic operating assumption, the radical separation of subject and object, us and them, humans and nature. Our exemptionist attitude is evident even in our language. The very term "environment" implies that which is external to what is assumed to be the central object of concern, human beings, and separates the really important stuff "in here" from everything else "out there." The Earth ceased to be a community to which humanity belonged and instead was seen as a commodity for use and possession, due to our mind's understanding of science. This human self-preoccupation ignored the reality that life and spirit are properties of the whole and its reciprocal interactions.

The power and expansion of human knowing through science has allowed us to reduce the material world into its component parts. This "mechanical" approach to nature refers to our idea that the world is assembled like a machine, acts like a machine, and thus can be treated like a machine, which has interchangeable parts. We take nature's kingdoms apart and isolate the components of these many living systems. Then, we rearrange them according to human logic which reflects that famous divide and conquer thinking. We keep only the parts we deem valuable for our needs as humans, which reflects our egocentric, consumeristic thinking. Then, we try to reassemble the retained pieces, like a machine, and expect the system to function as before. Sort of like taking away one's heart and still expecting to hear a heart beat. As it takes our heart, veins, lungs, brain and other systems in our body working together for us to breathe and live, it also takes the many systems, all those little kingdoms, in nature working together to make them a living ecosystem that will function properly.

An even scarier thought…if we view things, including nature, only as machines, what happens when we leave out a part, simplifying the "machine?" A helicopter crashed in Nepal some years ago, killing two people. A helicopter has a large number of parts. The particular problem here was with the engine, which is held together by nuts and bolts. Each bolt has a thin wire that runs through a hole in its end, to hold the nut in place once it is screwed on the bolt. This particular helicopter crashed because a mechanic forgot to replace one of these thin wires when repairing the lateral control assembly unit. The nut on the bolt missing the safety wire vibrated off its bolt, the helicopter became unstable, and the pilot lost control. All this was caused by one little piece of wire that was missing, altering the entire functional dynamics of the helicopter.[140]

Nature is no longer a mystery. An eclipse of the sun or full moon hold no more magic than any other day. We now believe in human will and independence. We command the world of nature. As we only focus on the well-being of humans, we only feel responsibility for parts of the natural environment that provide direct material assistance to our society, such as food, energy, materials, and waste disposal. Within this scenario, there would be no logical reason to feel responsible for the fate of species, lands, and those smaller kingdoms that invisibly to our eye support ecological functions.

Science can be viewed as humanity's collective ego asserting human will, creativity, and independence, breaking the limitations and superstitions that bounded us in previous ages, and now penetrating and commanding the world of nature that previously encompassed us. To continue to assert the extreme degree of independence and "false sense of omnipotence" given us by our mastery of nature, however, now threatens to destroy all life. Our evolutionary imperative, as in the Copernican revolution, is to leave this adolescent phase and progress to a more mature understanding of our true relationship with nature.

But, not everybody is immediately able to see all the connections that abound in nature, which is fine as we all are not going to understand everything. But, as long as we don't get caught up with divide and conquer games, one can see how two or three parts of these kingdoms might function together and then gradually expand their understanding as they need. People can learn what life supports what life, or what birds eat what bugs, all as part of what we have come to know as the very large "web of life," so then we can find ways to bypass harmful pesticides, for example. They can also then begin to figure out where their water comes from and who's farm it travels through, and what kind of bacteria it picks up if it flows near a factory farm where open lagoons can contaminate it? Or, what kind of plant, fish, or microscopic life is needed to naturally purify the water instead of having to use tons of chemicals?

8 The web of life in nature is as detailed, complex, and sensitive as a spider web you would find woven between two trees in the woods, as this one on Virginia's Eastern Shore.

Relationship among things is the essence of sustainability as well as the common strand in the web of life. Relationships take on many forms: intrapersonal; interpersonal; or between people and the environment as well as between people in the present and those in the future. The truth of all relationships and accepting the reality there in is the beginning to understanding. The successful efforts in sustainable development consider a dynamic and evolving set of interlocking systems—environmental, social, and economic—that interact and change. And, the conservation of natural resources is the fundamental issue because we are not absolutely sure what is connected to what. It is the open door to economic and political progress. Conservation of natural resources is the **only** door to progress.

THE GREATEST ILLUSION: MIND-HEART SEPARATION

So why are we not more sensitive to these many relationships within the natural world? Most of the socioeconomic institutions of modern industrial societies are based on the pursuit of material progress through separation from and conquest of nature, or in other words, we are gainfully pushed into capitalistic ways and views. If it doesn't bring a profit it is not of value. There is a resource piracy that has become the basis of economic growth, which, in turn, has become the dominant measure of social advance. The limits of this materialistic philosophy are now clearly characterized in the accelerating destruction of Earth's ecological systems.

A main part of this dilemma is the illusion that we can exist outside of nature, which is a conditioned response that generation after generation has come to believe. A thinking pattern that was not established overnight—a pattern that somehow the environment, these ecosystems, are separate entities from humans. Somewhere we have learned to view nature as materials and parts, just as we would view a house. Looking at how it is constructed, what kind of design it has, and what kinds of materials are used.

This mechanistic world view is quite incorrect. A house is not a home! For example, when we envision home, not just the house, we imagine not only the house structure but the relationships within that structure—the relationships among people in the house and between people and the environment of the house. This view goes beyond the physical structure, that which we envision in our minds from our knowledge, to embrace the feelings of the heart.

If there is a moral absolute, it is that desecration of the natural environment, risk to animals, plants, and humans when land, air, and water are fouled by others, is wrong. Is our mind telling us this, or is our heart telling us what is right and wrong?

Our minds and knowledge are influenced by science. Our hearts on the other hand are influenced by many things and sometimes may even offer some form of moral guidance. The human-nature split, which is the legacy of the "technology and science can fix all" (mechanistic world view), is directly related to this division of our minds from our hearts. Our mind's understanding of science, not our heart's moral beliefs, seems to be what is leading us astray…a loss of heart that has led to treating nature as a commodity from which we are independent and separate.

By separating ourselves from nature, we have justified our trying to control the uncontrollable. We have initiated a self-destruct format that can only be remedied by a change in mindset by the individual, in that we begin to listen to our hearts as well. Solving the ecological crisis, saving the kingdoms that support all life, will indeed require some support from our more compassionate side. Thus the ultimate answer—elect a woman president!

Science is a great way to express human intelligence, and is valuable as an organized approach to develop knowledge. The insights and skills that represent scientific accomplishment, however, must look to the force of spiritual awareness and moral principle to ensure their appropriate application. Dialogue between scientists and moralists can facilitate the standing of humanity within the bounds of nature throughout the world.

The greatest triumphs in science are not, after all, triumphs of facts but rather triumphs of new ways of seeing, of thinking, of perceiving, and of asking questions.[141] These ways of seeing and thinking are strongly influenced by what our hearts feel. The reality that is investigated by science, through experimentation and reasoning, and that is illuminated through the progressively revealed truth of moral beliefs then, ultimately becomes one.

THE ETHICS OF OUR FAILURE

A lack of earthly resources is beginning to cause us to face the ethical dilemma of our failure to meet the basic needs of more than a quarter of the Earth's present population. The U.S. has less than five percent of the world's population, yet uses 25-35 percent of the world's resources and produces 25 percent of the world's pollution and trash.[142] Shortage of natural goods will affect weaker members of society, leading to social disorder. Although post-Cold War conflicts such as in Haiti, Somalia, Sudan, and Rwanda are characterized in part by ethnic differences, territorial overcrowding and food shortages are contributing factors, providing early warning of the avalanche of global environmental problems related to disturbing Earth's delicate web and exceeding its capacity to provide natural resources for everybody. Our backyards have become very small.

We have already begun observing subtly the implications from increasing water shortages with regard to disease, global hunger, civil unrest, and even war.[143] Access to clean water can mean the difference between health and disease for millions of children. The number of people without access to safe drinking water could jump from a present 1.4 billion to 2.3 billion in 2025 unless present water shortages around the world are addressed. This could, and already is in some cases, leading to increased nation conflict.

Rain fed agriculture will not be able to keep pace with the world's growing population. About 80 percent of all water used each year goes to agriculture, and demand is increasing. Therefore, in this century about 60% of the increased food production needed will have to be supported from irrigation using surface waters and wells. But these sources are already in trouble.

Water scarcity is becoming the single greatest threat to human health, the environment, and the global food supply. China, for example, is facing an impending water shortage that has the potential to undermine its need for massive food production, boosting world grain prices, and precipitating political instability in many countries.[144] Provinces in Pakistan, and between Pakistan, India, and Bangladesh, have feuded for years over water allocations from common sources because of rise in human and livestock populations, an increase in cultivated land, and a change in cropping patterns.[145] Rivers between the countries are a constant source of tension.

Skirmishes in the 1967 Arab-Israeli war were fought over water diversion projects on the Jordan River. Has this changed since 1967? Most would think that many of the struggles among Israel and its neighbors is over land now. Well its not as much over land as over water.[146] Likewise, many believe the next war in northeastern Africa will be over the waters of the Nile, not politics.

Nature is collapsing on a global scale for the first time ever.

Chapter 7

Traditional Economics

"The GNP does not include the beauty of our poetry or the strength of our marriages, the intelligence of our public debate or the integrity of our public officials. It allows neither for the justice in our courts, nor for the justness of our dealings with each other. It counts air pollution and cigarette advertising and ambulances to clear our highways of carnage, yet it does not allow for the health of our children, the quality of their education, or the joy of their play."

Robert Kennedy

For 400 years persistent growth has been the central organizing principle of society in general.[147] Continued, uncontrolled economic growth seems the norm in light of the fact that in the last century there has grown such a dependence on economic activity. Think about all the emphasis placed by Wall Street on the desire for a continued growing economy in the beginning of the 21st century. More and more demands are now being made on human and natural resources, as well as more demands on physical and financial capital from our growing economy.

THE INDUSTRIAL AGE

The Industrial Revolution planted the seeds of what is "now" in both landscape and mentality. Many of the basic intentions behind the Industrial Revolution were good ones, which most of us would probably like to see continued today: to bring more goods and services to larger numbers of people; to raise standards of living; and to give people more choices and opportunity, etc. But there were crucial omissions. The great thinkers omitted preserving the diversity and vitality of forests, rivers, oceans, air, soil, animals, people and their cultures. This preservation was not only missing from their plan but also they were very unaware of the avalanche of problems this vital omission would ignite.

Our economy has reached the scale where it can and has caused the extinction of whole populations of fish, clear-cut forests, polluted many waters, dirtied the global atmosphere, and caused a thinking that even people are a throw-away commodity. While successful in providing goods, services, and jobs for some, the economy is still dependent upon environment and diversity. Basically what has happened, and is happening, is industrialization is destroying the very natural resources that are required to support it. And if that is not enough, society is now learning that as much as 40% of world deaths may be due to environmental degradation caused by growing economies.[148]

But how has this happened when economic markets were established to serve people? Why else would markets exist. And ultimately they cannot exist without people, without the consumer. How efficiently markets serve the people is the measure of their legitimacy. How an economy serves all segments of society is what economy's performance should be judged by. How well do markets and economies aid those factory workers, or kindergarten teachers, or the homeless and desolate? That is the question that seems to be the reality of an economy's performance. Yet, today's markets and economies leave in their path millions and millions of refugees. What does that say?

In reality economics have eliminated the human component. Having eliminated humans, economists then eliminated the behavior and attitudes of humans. Finding the interactions among people too hopelessly complex and difficult to measure, economists choose to observe the behaviors and trends of markets rather than people. Market behavior involves prices and flows of money, which are easily observed and measured,[149] and just as easily masked and disguised. A Catch-22 exists here; what brought us more comforts is now a burden on our planet that threatens the survival of all human civilizations. Thus, we now find ourselves in a position to begin questioning our economics. An economics that eventually will destroy us is not an economy that works for the people.

In Morocco for example, high population growth and corresponding economic growth rates are bringing significant pressure to bear on natural resource and other multi-faceted environmental degradation.[150] In Pakistan massive economic growth has resulted in significant environmental damage throughout the country.[151] Pollution is playing havoc in coastal waters, with the scale of its effect on marine, bird, wildlife, and human health alarming. Various illnesses are on the rise because of lead, carbon dioxide, sulfur dioxide, and other substances in the air of large cities. From another part of the world, 9 out of 10 of the most polluted global cities are found in China.[152] Counting the costs of worker's sick days and health care from pollution-induced diseases, as well as the amount of air-filtering forests and farmland lost to environmental degradation, the price is high to China's further economic growth…as high as 8% of their total Gross Domestic Product (GDP).

Another way of looking at the effects of pollution on economies is to determine what fixing environmental ills means to our prosperity. The U.S. Environmental Protection Agency (EPA) has estimated that as a result of the 1990 Clean Air Act and what it requires in environmental remediation, benefits from less premature deaths, less health care from asthma attacks, and lost work from incidences of bronchitis or other pollution-related illness, showed a savings of $110 billion.[153] By contrast, the costs of achieving these health and ecological benefits imposed by the Clean Air Act on industry for example, were about $27 billion, less than one fourth the economic value derived from the benefits of less pollution.

But, there seems to be a dilemma in the need to stimulate the economy to make jobs, so everybody has a chance to prosper, which is opposed to the need to slow the economy to avoid upsetting nature's balance. For example, federal reserve policy is being used to slow the economy, while congressional tax/borrow and spend is being used to stimulate the economy. It's like driving with the brakes and the accelerator pressed at the same time.

Why do we have so much difficulty imagining a no growth economy? Could the reason be that a steady-state economy, one that does not keep growing so that we can then begin protecting the environment, will force us to confront the major problem of inequality, poverty? If you don't think so, consider the fact that the economic market system has no ability to distribute goods and services evenly or fairly. We have always been able to argue that poor people will be better off next year because their slim piece of the pie will grow a bit larger as the total pie expands from our continued economic growth.[154] But if economic growth is constrained by concern for increasing environmental problems, the benefits of development that trickle down to the poor gets smaller, and with a cessation of growth virtually disappear.

Since people's demands for economic betterment are not likely to disappear, once the pie stops growing, the poor will begin making demands for equality and re-distribution of benefits through social unrest. Today's public notice that this progress is no longer working for the masses leads to a new revolution—a revolution of efficiency, effectiveness, equality, and environmental soundness.

This revolution must recognize the basic elements of the problem. The total throughput of human economy must be kept small enough to avoid exceeding two physical limits of the ecosystem: its capacity to regenerate or refuel itself (its resources) and its ability to absorb wastes.[155] To lessen this risk, at some point externalities of an economy that represent environmental costs must be internalized into the dynamics of economic activity, truly reflecting these costs.

For example, the chemical industry may have already passed the point at which its toxic discharges are costing society more than the benefits provided by its products (*e.g.*, the "plastics" commercial on TV and all its rewards).[156] By not showing the **real** costs of these insults on the environment, we are kept from linking our economic decisions to our concern for healthy environments and the appropriate decisions to guarantee their health and sustainability. In essence, we must fully accept the fact that economic development is closely linked to natural resource stewardship.

Consider our fascination with the automobile. We are all aware of the chemical air pollution associated with the gasoline that fuels our cars. And yet, the subsidies provided by the U.S. government to keep gasoline production costs, and thus consumer prices, well below the world average (*e.g.*, $1.60/gal average in the U.S. compared to $3.50-4.00/gal in Europe and Japan) only serves to stimulate reliance on the automobile by U.S. consumers. The result...more miles driven and more air pollution in U.S. cities. In addition, people also live further away from where they work and shop, increasing the sprawl associated with many cities and consequently, growing social as well as environmental problems. So our economies are not truly showing real costs for the things that consumers enjoy, and even cherish, which thus has a "domino effect" into all parts of our social and natural environment.

Beyond the effect on natural resources, another cost of our economic approach for employing too much human capital and technology involves the effect on everybody having a job available to them. Most would agree we should manage the economy to do more than just providing goods and services...to also provide enough jobs to go around. Industrial growth therefore means an increase in the number of jobs, but not necessarily the quality of jobs. And it is the quality, not the quantity of jobs, that should really be the focus as a measure of economic health.

But creating enough good jobs to go around has been made more difficult by the trend toward the use of labor saving technology. Is there equality in automation? The market insures that producers will either use the latest labor saving automation to cut labor costs or go out of business. Then, to add fuel to the fire, we have learned from experience that automation will cause unemployment if we don't consume more...so we can produce more.

Our adoption of automation and the corresponding need for growth to support this automation, but also to fuel additional economies supporting those unemployed by automation, have made the consumer economy not a luxury but rather a necessity, and also a **vicious circle**. Since the industrial revolution, economic growth has been the key to providing enough jobs. Growth in economies has compensated for the loss of jobs from the increased use of labor saving machines and computers, by stimulating increased business, manufacturing, and resulting consumerism to further enhance the job base. Our history of continued economic growth, fueled by increased consumerism, is the only explanation for why machines haven't caused greater unemployment.

Let's examine more closely relationships between economic growth and more jobs. Smart growth is often equated to increased jobs for local people. But can you also imagine that creating jobs in a community of 10,000 people able to work might actually increase the number of people out of work?[157] If a community's unemployment rate is 6% (600 people) and a new factory is recruited to locate in town, the promised hiring by the new factory could lower the unemployment rate to 4% (400 people), initially. But because of increased community prosperity new people move into town, causing a population increase to 12,000, and restoring the unemployment rate to 6%. Now, however, this unemployment rate is 6% of a larger population than before the factory located in town, so the total number of unemployed has increased to 720 people.

The top-dowess of industrial enterprise has made us smarter than the average bear. Modern business is propelled by the notion that human intelligence can transcend material limits and thus, serve the powerful economic elite. Introduced technology, which replaces traditional human labor without removing the people from the equation (labor pool), can cause an internally sustainable social/environmental system to collapse.

For example, prior to importing the technology of clear-cutting, the forest of Cergov was selectively logged with horses.[158] In addition, horse logging had been biologically sustainable for centuries, as were the economies of the small mountain villages located in the upper valleys near the edge of the forest. Now the jobs once sustained by horse logging are gone, and the topsoil of the forest has been eroded away, all because the technology of clear-cutting forests has replaced a time-honored profession of sustainable timber harvest. The people who once made their living from the forests and agriculture on productive lands, must now commute to the cities to find work, and the villages have lost part of their cultural heritage.

9 *Excessive logging, environmentally unsound agricultural practices, and other poorly designed building modifications to the landscape can create serious risks for erosion of land to occur.*

INDEPENDENCE AND VICIOUS CIRCLES

Human economies digress from the successful paths of nature when we get caught up in a vicious circle. Vicious circles are dead ends, because instead of correcting for an instability reported by feedbacks, they only intensify it,[159] as illustrated by the desire to create more business, jobs and consumption of goods to sustain those workers laid off by improved technologies. The collapse of the Grand Banks fisheries off the northern New England coast is a good example. New England fisherman ignored the feedback messages from dwindling fish supplies. Instead of lessening the fishing pressure they responded by investing in larger ships and bigger nets that could go further out to sea. The problem wasn't the feedback mechanism but rather the response to it. By increasing the efficiency of the industry, a collapse in both the fishery stocks and the fishery economy eventually occurred. The correct response would have been to let up on the fish catches when they started to diminish.

In a similar fashion, during the rush hour in crowded suburbia many roads are congested that cause traffic jams and longer travel times. In most cases the response is that streets need to be widened and more roads built because existing roads are inadequate. But because bigger and better roads and faster speeds encourage more use of cars and also greater numbers of cars from more people moving to the suburbs, congestion builds up again, requiring more road expansion. Is this not a vicious circle? The alternative might have been that cars and trucks were being depended upon too heavily, to the detriment of optional forms of transport, or that better zoning regulations were required that did not so greatly separate everyday conveniences and work from residences.

In nature, vicious circles are damaging but self-terminating.[160] That is because those natural players involved will in one way or another eventually respond to the many feedback messages they are receiving, travel a different fork in the road presented by coincidences that occur, or their populations and communities will collapse (*i.e.* extinction).

Not so with the human economic system. There will always be the search for a "fix" with a new technology to get us out of the trouble we are in. But can this go on indefinitely?

INABILITY OF MARKET SYSTEMS

The complexity of our modern economic systems cause us to sometimes ask "when is it that fair is fair versus fair is feudal? Every economy faces 3 problems: allocation, distribution, and scale.[161] What do these terms mean and how do they muddy the waters toward achieving potential solutions?

Allocation refers to the circulating of resources among different products—in other words, deciding whether we should produce more corn, more cars, more bicycles, more jelly beans, or more hospitals. We can't have everything, so we must allocate our resources in some way to provide the goods that people want most and can afford to pay for. The way we do this is "the market" which sets relative prices for goods.[162] Prices act as signals or feedback mechanisms that cause people to put more (or fewer) resources into creating particular products, indicating what other people are willing and able to buy.

High prices for goods stimulate increased production of those goods while low prices depress production, bringing supplies in closer relation to demand. Price feedback by itself is inherently well integrated and under normal circumstances should work effectively in the appropriate allocation of resources.

But two difficulties can easily arise. First, from the production side high prices will stimulate the substitution of goods in short supply, possibly goods not previously in existence, or the use of a resource that could eventually become depleted.[163] For example, in 15th century Europe the demand for books exceeded the supply of individually, hand copied books. Thus, printing with moveable type presses became economically feasible. Printed books using paper from trees became much more plentiful. You can image the rest...more paper needed, more trees cut.

Second, from the consumer side, the data that prices convey regarding supply and demand must not be sloppy or ambiguous, but rather true. The data on prices, however, can be false, where costs, a major ingredient of prices can be falsified, so that prices are also then falsified. How is this done? The use of subsidies, usually financial incentives from governments, can supplement production costs, keeping them lower than if production was done without federal assistance. Likewise, taxes can favor some types of investment and production of goods, while penalizing other types.

The second problem faced by every economy is distribution—apportioning goods (and the resources they embody) among different people, not among different products. Goods should be distributed in a way that is fair. If you don't believe this statement is true, think of an extreme case. If one person received 99% of all the benefits provided by the U.S. economy, and all other citizens had to divvy up the remaining 1%, almost everyone would agree that this was an "unfair" or unsatisfactory distribution of benefits that showed real inequality for the majority of citizens. These people would say, "there is something wrong with this picture." This extreme example is intended to show that nearly everyone agrees there are "fair" and "unfair" distributions of goods.

Conflict is born in the presence of this inequality. If a large segment of society is cut off from the benefits of the economy, this breeds envy, distrust, animosity, and ultimately fear and danger for everyone. It weakens the fabric that makes one out of many. The fact that there are 25 million environmental refugees reflects the inability of the current market to distribute goods and services evenly or fairly. Left alone a market economy tends to create inequalities that grow larger as time passes; tends to make the rich richer and the poor poorer.[164]

The market cannot solve the problem of unfair distribution. And current economics will not stop the rape of the world. The problem of how natural resources are used and distributed must be solved by people deciding what is fair. Fairness is morally and ethically right. It will bring the greatest good to the greatest number, and because it is the only way to preserve our most important ideal—our democracy...without it we will surely lose our liberty. Then comes the making of public policies intended to achieve a fair distribution. After those decisions have been made, the market can allocate resources efficiently within the politically-established framework of fairness. In the end we have a political system less tied into lobbyist concerns and re-acquainted with the heart of the people. A political system that once again speaks of liberty and justice for all.

The third economic problem is one of scale—how large can an economy become before it begins to harm the ecosystem that underlies and sustains it?[165] Should the economy be global, national, regional, or local in scale (size)? Here again, the market does not, and cannot provide answers. The market offers no mechanism for deciding what is a desirable scale or a means for achieving that scale. And in fact, in many instances flawed assumptions and extrapolations by economists through history have even suggested that the economic specialization of regions and nations is the most favorable path to more stable and prosperous economies globally.[166] This is the justification for global scale economies which immediately put local and regional economies at the mercy of changing variables totally out of their control and also usually places undue pressure on natural resources that support the specialized economy, the so-called "only business in town."

You can have an efficient allocation of resources and a just distribution of benefits, yet still have an economy that grows too large and consequently damages the ecosystem. The "take home message" here is that each of the above three problems—allocation, distribution, and scale—is separate and each must be solved separately. But, all should be considered with an eye on the ultimate effect on natural resources from potential solutions.

Unlike in economics and the working of markets, nothing in the natural world operates to maximize present value to the exclusion of all else. Nature, like mathematics, is pure and uncorrupted. It is governed by its own laws, which people might learn but never master, nor bend nor twist to suit themselves. In contrast, human laws of economics and jurisprudence are often not in harmony with the Earthly laws of biology, chemistry, and physics. Thus, competition is usually the mode of operation, as well as the killer here.

Competition becomes the key issue with regard to most natural resources considered in an economic light, because in most cases these resources are often considered "common-pool" resources, available to everybody or to many, but at the same time usually outside of the market system. Have you ever heard of the "tragedy of the commons?" What makes the stewardship of this kind of resource most difficult is the diversity of user attitudes toward the resource. As long as competition and reliance on market signals, which as we have already observed can be false, are the overriding principles of our economic system, we can only destroy our environment because it has become the battlefield in which the war of competition is fought. There are no signals of trouble, like increasing prices of environmental resources available to everyone, until it is too late.

With this many different views on life for people who have equal access to any natural resource (think about gold), you can imagine the conflict that eventually causes severe competition. The larger and more immediate the prospects for material gain, the greater the political power used to ensure and expedite exploitation, because not to exploit is perceived as losing an opportunity to someone else. And it is this notion of loss that we fight so hard to avoid. In this sense, it is more appropriate to think of resources as managing humans than of humans managing resources.

Shifting from a posture of competition (winner and looser) to one of seeing that everybody is a winner (win-win), is the way to proceed. Individuals overcoming their tendency to evaluate their own benefits and costs will be the beginning. Then individuals shifting to look more intensely at the total benefits and costs for a group. What about the alternative of moving beyond the traditional market system approach and integrating economic benefits with environmental protection?

Consider the state of salmon in the Northwest rivers of the U.S. Studies suggest that efforts to improve stream health, the habitat of these salmon in juvenile and reproductive stages of their life cycle, will also improve the environmental efficiency of our economy, which will translate to cost savings and increased productivity for firms, communities, and the public sector.[167] This directly challenges the notion that saving Pacific Northwest salmon is a snag to an area's economy. Figures provided by many businesses employing one third of the work force in Washington and Oregon (USA), ranging from small restaurants to The Boeing Company, show that improvements in streams and salmon habitat also benefit the bottom lines of these businesses. Actions to protect water quality generated significant cost savings over environmental cleanup costs, increased efficiency, and raised productivity for firms, organizations, and the region as a whole.

THE PROBLEM OF ALL GOODS AND NO BADS IN ECONOMICS

Numbers are often used to tell us if we are doing good. We add up all the goods and services to see how well any particular country is doing.[168] All this adding up is referred to as either the Gross National Product or Gross Domestic Product. Gross National Product (GNP) refers to the total amount of money the economy of any nation is worth on an annual basis. It essentially counts all the income of business. If GNP is rising, most would assume our standard of living is rising too. There is also the Gross Domestic Product (GDP) which tells us the number of all the goods and services a country is gaining income from. So once all these goods and services are added together and made one, the GDP or GNP is our measuring stick. Making it look good is a priority.[169] In the U.S. we are doing really, really good.

But this measuring stick excludes such things as divorce lawyers and security systems. Unfortunately, the GDP does not subtract from its total number the ills in society or the damage that we might see, for example to the environment. If I buy a car it is a plus for the GDP. When I junk that car, the service of getting rid of the waste (non-recyclable parts), the service of storing its toxic parts, and the service of recycling some of it, are all a plus for the GDP. The same goes for TVs, appliances, clothes, bottled water, etc. But, the GDP does not account for the release of toxic substances from the junked car into our environment, for the broken home created by divorce, or for the burglary loss that prompted the security system to be installed in the first place.

Paul Hawken said that the most damaging aspect of the present economic system is that the expense of destroying the Earth is largely absent from the prices set in the marketplace.[170] The damage to the environment after it has been stripped, cut, burned, or spilled upon is not counted in the GDP. Thus, we have no way of measuring whether our economy has passed the point at which costs have begun to exceed benefits because we count all production of goods and services as "good."

While we focus on earning our living, we tend to ignore what we have been given by nature for no payment. Air, water, and other essentials of life provided freely by nature are treated as valueless...that is, until scarcity and privatization render them marketable.[171] While price reflects labor, the costs of producing goods and services ignore nature's resource inputs. Market prices generally say nothing about the size of remaining natural capital stocks or whether there is some critical minimal stock size below which recovery is impossible. Think again of the North Atlantic fisheries story. Subsidies, low fuel costs, and high-tech factory freezer-trawlers enabled industrial fishers to access previously unreachable stocks of North Atlantic ground fish. This maintained market supply (and relatively low prices) even as the stocks were being depleted.

In tallying up GDP chemicals are also counted as "good" and the products they allow us to make are counted as "good." This makes sense. But when our chemical factories produce chemical waste dumps that must be cleaned up at huge public expense, those costs are counted as "good" too, instead of being subtracted as "bad." If children get cancer from exposure to chemical wastes (*e.g.*, the movie "A Civil Action"), their hospitalization, their radiation treatments, their chemotherapy, and their funeral expenses are all counted as "good."

10   When an oil spill occurs, such as the IXTOC I oil well blowout in the Gulf of Mexico in 1979 (above), the costs for environmental cleanup and remediation from this damage are counted as an economic good (on the income side) instead of an economic bad (on the cost side) as far as the Gross Domestic Product is concerned. The lighter color on the water's surface in the photo is the oil spewing from the well blow out located in the middle of the photo.

A perfect example is the following. In the early 1990s the Exxon Valdez oil tanker ran aground in Prince William Sound, Alaska. The millions of gallons of spilled oil killed millions of animals and cost millions of dollars to clean up. The jobs created and materials manufactured related to clean-up activities of the polluted water and beaches, as well as the aid provided to impacted communities, made the U.S. GDP go up. In this case, using the GDP as an indicator suggests that we should get more oil tankers to run into rocks more often.

As preposterous as it may sound, most nations, including the U.S., presently treat depletion and damage of their natural capital as if it were income—a major accounting error. What is being depleted are our valuable natural resources and the parts of the Earth system needed to adequately treat our wastes. Depletion should never be treated as income. Depletion is a cost, not a benefit.

CONSUMERISM: THE SECOND HALF OF THE POPULATION PROBLEM

The seven most harmful consumer purchase activities could include the following:[172]
- cars and trucks;
- meat;
- fruits, vegetables, and grains;
- home heating and cooling;
- household appliances;
- home construction; and
- home water and sewage.

Environmentally speaking, the worse thing you can do is drive your sport utility vehicle to the steak house for a prime sirloin. It comes as no startling revelation that automobiles are at the top of the list of environmentally damaging products. But what is surprising is that meat comes in at number two. The industrial production of beef, poultry, and pork pollutes waterways and air, fouls the land, and gobbles up valuable land and other resources.

In terms of water pollution, beef production is 17 times more damaging than all the contents that go into making pasta. The water pollution from manure, as well as the amount of electrical energy, fuel, fertilizer, and pesticides needed to raise cattle is significant. With the many cows produced and products manufactured to feed them, there is a lot of manure...2.7 million pounds per year. This manure causes everything from massive fish kills to water contamination. But for some reason this industry is given major tax breaks encouraging them to move to depressed rural areas to increase an area's economy. Beef production is also 20 times more damaging to wildlife habitat than pasta production, because it uses more land.

Vegetables, fruits, and grains are third on the list because their cultivation usually entails large quantities of pesticides, herbicides, artificial fertilizers, and irrigation water. In 1945, when all agriculture was essentially organic, crop loss due to insects was 7% per year. Now, after a 1,000 fold increase in pesticide use, losses are on the average 12% per year.[173]

The family who is going out for steak, is probably also coating their yard with fertilizers and weed killers for greener grass. This same family probably also throws away 25% of the food they purchase and cannot even imagine a diet with a reduction of meat consumption. They also leave lights on everywhere, have multiple TV sets, probably drive 2 blocks down to see a neighbor rather than walk, and consistently forget to recycle.

This same family may build a house at some point, disregarding the use of building options that represent environmental friendly housing due to a small up front increase in building costs. But the benefits would include the water being pumped with a windmill, and the home heated with solar panels. This same family is of the thinking that they do the 9-5 jobs 5 days a week and deserve luxuries on the weekends.

The belief that this family lives as every other family does in their small suburbia, validates that there is nothing wrong with their lifestyle. Multiply that by 300 million, reinforce it with advertising, and we have a picture of western civilization in the 21st century. Until one day, one member, perhaps the mother of this family, discovers she has cancer. This cancer could be attributed to anything from the stress of keeping up with the neighbors to too much DDT tainted fish, the hormones in the red meat, or the dirty air and water that subtly poisons. But the concrete experience for this family, aside from the emotional trauma of illness, is the loss of one income.

When one's very breathe is threatened due to everything one has turned a blind eye to; when reality comes smashing through the french doors; this family may wake up. This family may realize that this cancer call is only the beginning of a series of wake up calls; wake up calls that blatantly reflect the reality of how suburbia feeds into today's toxic air, water, and food. This family represents the population of the industrialized world and it is easy to see why we are in the predicament we are in. These families are not bad, they simply do not understand the power of the masses: their power in creating destruction and their power in initiating change. Their status as environmental victim is only a few inches away from those who are environmental refugees.

AN ECOLOGICAL ECONOMICS

We as human beings cannot continue to withstand this consumer style invasion on the environment. Limits to growth from such things as natural resources are invisible to static monetary analyses because monetary expansion itself is not bound by physical limits. But, a growing ecological perspective seriously challenges this money-based view. Clearly the physical consumption of natural income by one person preempts any other person from using those same income flows. If global carrying capacity has been exceeded, then consumption by the rich of limited natural resources already undermines prospects for the poor in benefiting from these same resources. Economists are finally beginning to argue, given the irreversibility of many kinds of environmental damage, that it is appropriate to maintain healthy environments (e.g., biological diversity) as an option.

Presently environmental components like biodiversity are not adequately protected, however, because their value is not included in the market signals that guide the economic decisions of producers and consumers, and thereby the overall operation of the economic system.

Biological diversity and its positive effects on environmental integrity (sustainability), is widely recognized as important, but how much money is it worth? Biological diversity is not sold around the corner or produced and bought regularly like a commodity. Environmental goods and services often have no market price tag and a considerable amount of uncertainty can surround their true value and significance.

In addition, the logic of market failure has led economists, and increasingly biologists as well, to argue that the critical environmental resources should be incorporated into the market system in some other way. The appropriate analogue to a traditional goods production, consumer market scenario, would be to maintain a safe minimum standard of natural resources, the setting of a lower limit on the quantity of a resource that must be maintained by the price escalating "out of sight" if that sustainable level of the resource is exceeded.

Better-safe-than-sorry reasoning is now leading to the introduction of concepts that adjust prices upward to reflect true environmental costs and help assure the conservation of valuable natural resources. Many are now beginning to believe that a healthy environment and a strong economy, with equitable access by all, can coexist. Conservatives may have to overcome their inordinate fear of environmentalism. And environmentalists in turn may have to overcome their inordinate fear of progress as measured by enhanced development, instead of growth—but it is beginning to happen.

Sustainable development implies managing economic systems in such a way as to guarantee future generations the opportunity to enjoy a level of well-being equivalent to or better than that experienced by those now alive. Can we move from this destructive economic system that ignores the inter-relatedness and interdependence of all aspects of life on the planet to one which acknowledges and builds upon it?

There are choices that can reduce the poisons that shake up more worlds than one would have ever guessed. We can use increased durability (quality rather quantity) to provide more goods-in-service without the demand for high resource consumption. Increased durability is a substitute for more production. If we build products that last a long time we can own a lot of wealth without high resource consumption. The real meaning of sustainable is to last over time. What else could be so effective?

We are squarely on the downside of nature's limits. We can no longer pretend that we can achieve the greatest good for the greatest number. Population increases coupled with increases in consumerism are no longer an option. As we are confronted with the limits of the planetary ecosystem, we are forced to ask: "how much good can we achieve, for how many people, for how long?" We can have "the greatest good for a sufficient number" or we can have "sufficient good for the greatest number" but the "greatest good for the greatest number" we cannot have.[174]

Herman Daly favors seeking "sufficient good for the greatest number"—meaning the greatest number of humans that can be supported year after year into the indefinite future. If one's goal is to maximize human welfare, this is the formula that does it.

CULTURAL REVIVAL TOWARDS LEAN, EFFICIENT ECONOMIES

In a world at its carrying capacity, decision-makers have an obligation to pursue only those technologies, development projects, and life improvement strategies that reduce society's total, future ecological footprint. Any option that does otherwise contributes to long-run instability and uncertainty in a negative-sum game that is detrimental to everyone. And as well, decisions that increase one group's over-consumption at the further expense of common-pool resources otherwise available to everyone, imposes dangerous instability in a societal system that can only effectively operate over the long-term through equity.

What is called for now is real economic and social development: qualitative improvement with no or minimal expansion in resource exploitation and explicit recognition of the inter-relatedness and interdependence of all aspects of life on the planet. We can move from an economics that ignores this interdependence to one which acknowledges and builds upon it, by developing an economics that is fundamentally ecological in its basic view of the problems that now face society. The economic system that we are entrenched in does not work for us. Consumerism does not work for us. Environmental victims and refugees are not just a temporary concept or passing global crisis. It's time to greet a new era where growth takes the back seat to development and people come before profits.

An intriguing thought involves trying to learn a new economics from the way nature functions.[175] Instead of our traditional approaches to advancing technologies, we could consider the idea of biomimicry, imitating the chemistry and biology dynamics of nature to produce materials and products by methods that are benign and produce wastes at the end of their lives that can be benignly returned to nature for degrading/decomposing.

In this sense biomimicry is a form of economic development. Nature affords foundations for economies and sets their possibilities and limits. All kinds of people are now coming to understand that their success depends on working knowledgeably with natural processes and principles. Ironically, the common Greek derivative, oiko (translated as eco), for both economy, meaning **house** management, and ecology, meaning **house** knowledge, suggests managing one's house knowledgeably, in harmony with the natural environment.

Thus, many people now engaged in economic activities do realize it is important to learn from nature and apply that knowledge to what we do.[176] Architects and engineers accept the reality of natural forces of tension and compression in their building. Wine makers, cheese makers, and bakers value cooperative relationships with yeasts and bacteria. Economic development is a matter of using the same universal principles that the rest of nature uses.[177] The alternative is not to develop some other way because some other way does not exist.

This is another way of saying that when species inhabit an ecosystem, and as populations and communities of plants and animals diversify (differentiate), the ecosystem and its biomass or worth grows. As this diversity occurs, the continued development, stability, and success of the natural ecosystem depends on the co-development of the different plants and animals, similar to the many relationships established in a complex rain forest food web. And it is the co-development of species that continue to fuel differentiation or further diversity.

For example, a river delta needs both water and mud to form. Neither by itself can produce a new river delta and each is the result of a co-development process.[178] Inventors who combined silicon chips with typewriter keyboards (the personal computer), or any other devices with different economic beginnings, are not imitating plant cells or animal organs (things), but rather they are imitating natural principles of development and co-development (processes of systems to yield things) for the evolution of new ways of conducting human life functions.

Economic development is thus a version of natural development. Plants and animals together compose an ecosystem, just as collections of businesses compose a community economy. These economies require diversity to expand, self-refueling to maintain themselves, stabilization through self-correction (forks in the road), and co-development to further progress, just as ecosystems in nature require the same processes. Although complex, all are brought into an order that makes sense through unpredictable self-organization of the system, whether economic or natural in character.

Most importantly, it is a misdiagnosis to say our problem is really economics. Our problems are moral, spiritual, philosophical, and behavioral. Many believe we can attract a majority of people to a new vision because, done right, the vast majority of people will benefit. This would include the following.

- A vision that makes business and profit-taking responsible: holding development to a standard of environmentally sustainable, requiring companies to make products and use processes that are good for the health of workers, consumers, and surrounding communities, and that restores rather than damages the environment.

- A vision of good jobs that can support a family and allow time for leisure, education, and social participation, balancing our work as part of our lives with the necessary evil of having to pay bills.
- A vision of economic development that is socially sustainable, with a goal of overcoming historic divisions and oppressions related to race, class, gender, and national origin: where there are no refugees, no less than, and no "not good enoughs".
- And finally, a vision that holds a strategic alliance between the labor movement and political, democratic, environmental, economic, new immigrant, and social organizations: just as these entities came together to protest the World Trade Organization (WTO) in Seattle in 1999, so can they move together to make a more sustainable future.

The consideration of sustainable lifestyles on an individual basis ultimately depends on reducing our demands. Decreasing our ecological demands, mainly by limiting the material and energy throughput of the human economy, will follow from a consistent, personal reduction. Remembering our starving brothers and sisters in humanity will unleash the energy to embrace the needed change. Improving the quality of life for the world's poor, by freeing up ecological kingdoms, insures a future for everyone. A future where environmental victims and environmental refugees are only folklore.

Chapter 8

Excuses for not Trying Sustainable Development

"In the name of modernity we are creating dysfunctional societies that are breeding pathological behavior. The threefold crisis of deepening poverty, environmental destruction, and social disintegration is basically a manifestation of this dysfunction."

David Korten

We have before us a mixture of generations who have no idea of simpler ways...a culture that believes buying is a past time and consumerism is a way of life. Powerlessness is rampant, and choice seems to be only a part of history. And, as this automatic pilot is seemingly out of control, we find ourselves about to run out of gas. This poorly guided trip back from industrialization has left us a bit confused with tendencies that will haunt our children. A lot of the confusion about conservation, the declining environment, the meaning of sustainability, and why it matters has slowed progress toward achieving it. In today's materialistic, growth-bound world, most believe the politically acceptable is ecologically disastrous, while the ecologically necessary is politically impossible. Developing sustainability strategies that are consistent with the ecological bottom-line therefore, depends on the convergence of ecological and political practicality. The demanding challenge is how do we make sure that we are not, and are children are not, going to find ourselves as environmental refugees with little hope of recreating better days?

BLURRING THE ISSUES: ARGUING OVER DETAILS

The deliberate blurring of issues, conflicting interests, opposing world views, incompatible analysis, rising material expectations, and fear of change have led to a disorienting array of interpretations of sustainability and how to achieve it. And to deepen the waters of the progress of clarity, there is an addiction that has taken hold as we crave more and more things to feel normal. And where addiction stands, denial is usually rampant and lethal.[179]

Our slow support for living a kinder, more sustainable life can be explained by a general lack of understanding. Uncomfortable with change, wanting to maintain the status quo, many quickly buy the corporate agenda, which is a huge barrier in itself. Current beliefs feed from a history of "the land of plenty" and are almost genetic at this point. Society's consumer attitude, from shopping tendencies, to day-to-day habits, are firmly entrenched in excess, then cultural addiction—when a pattern is continued even in light of causing harm to oneself or others.

The commonality of excess also feeds cultural addiction. Just like an alcoholic hangs with drinking buddies, we believe our consumption is okay because everyone else is doing it. Corporations mesmerize us with 6 second flashes that own our thinking. We act out of fear of not having enough. Just as addicts protect their stash, so we stash and store until we have more things than we could ever possibly use. We respond from a place of naive ignorance as to the consequences of our actions. We are not bad, we do not act out of intentional malice, but our patterns are just as destructive regardless of intent.

Likewise, communities and their organizations who have vested interests tend to hide by distorting information. Such distortions do not depend on deliberate falsifications by individuals. Instead, people who are competent, hard-working, and honest can sustain systematic distortions by merely carrying out their roles on the job. Due to consumption, they work harder and harder for that new car or bigger house, rather than simply saying enough is enough and making due. The influence of the undeniable realities are leading to catastrophic failures. Just as an addict's culmination of their addictions hold potential for catastrophic consequences.

Progress on sustainability is slowed by scientists arguing over the details. The scientific world is composed of a bunch of individual papers, experiments, and opinions. That's the way we scientists work…on small problems, little tweaks here and there, which are fascinating and challenging in themselves.[180] The rest of it is not my problem!

Protecting the environment, however, is being held back by arguments within the scientific community about peripheral details. The simple analogy is that scientists might argue about the cause of withering leaves on a dying tree, instead of paying attention to facts they can agree on, such as the tree is in fact dying! In the midst of all this chatter about the leaves, we have not been paying attention to the "environment's" trunk and branches. They are deteriorating as a result of processes about which there is little or no controversy. In contrast, the thousands of individual problems (the leaves) that are the subject of so much debate are, in fact, manifestations, or simply symptoms, of systemic errors that are undermining the foundations (the tree itself) of human society. But if one begins with the trunk or branches, the answers become clearer and more consistent and just maybe the tree can be saved.[181]

If a politician or government regulator were to ask a random selection of scientists whether or not the reproductive organs of seals are destroyed by the chemical PCB, it is very unlikely that he would get the kinds of answers that would be helpful in trying to make a decision on protecting the environment from this chemical. **Is PCB a naturally occurring substance?** No, all scientists can agree that it is an artificially made substance. **Are PCBs chemically stable or do they quickly degrade into harmless substances**? All scientist could agree that PCBs are stable and persistent. **Does this chemical accumulate in organisms**? Most would agree that it does from numerous measures of wildlife over the years. **Is it possible to predict the tolerance limits of such a stable, unnatural chemical such as PCBs**? Scientists would absolutely say NO. **Can we continue to introduce such chemicals into our ecosystems**? Most scientists would say No, not if we want to survive, because we are not able to predict the tolerance limits of these chemicals by life processes and do not know when we have surpassed these limits.

The final answer is what the politician wanted to know from the beginning, since he is probably not really interested in the reproductive organs of seals or the exact details of PCB occurrence in the environment. Yet, most public environmental debate is preoccupied with such relatively minor details or in asking the wrong question to begin with. This happens whenever we fail to proceed from a basic frame of reference which makes it possible to focus on the fundamental issues without getting lost in a confusion of isolated details.

And then there is the old caveat, "more research is needed." In order to predict that you will die if you jump off the top of the Leaning Tower of Pisa, it is not necessary to calculate that it is 800 feet high at 20 degrees centigrade. In most cases we already know enough to take some action on environmental issues. The longer we delay, the more painful the sacrifices will be.

THE ROAD TO HAPPINESS

Our culture tells us that money, prestige, and public recognition are the goals we should be striving for, and that achieving them will give us happiness. Does the belief that a little more money would make us a little happier and that it is very important to be well off, after four decades of consistent rising affluence, really make us any happier?[182]

Since 1957, the number of Americans who say they are "very happy" has declined from 35 to 32 percent, not increased as one would expect from our growing economy. Meanwhile, the divorce rate has doubled, the teen suicide rate has nearly tripled, the violent crime rate has nearly quadrupled, and more people than ever are taking antidepressants! Ironically, once material sufficiency is secured people's happiness is usually no longer correlated with national or personal income.

An impressive body of evidence now suggests that people's satisfaction simply is not for sale. Eighty-eight (88) percent of people say our society is just too materialistic.[183] Not only does having more things prove to be unfulfilling, but for people whom affluence is a priority in life tend to experience an unusual degree of anxiety and depression, as well as a lower overall level of well-being.[184] Americans for example, are encouraged to strike it rich, but the more we seek satisfaction in material goods, the less we find them satisfying. The satisfaction has a short half-life.

Most often we as consumers actually don't feel we enjoy more fulfilled lives once our basic needs are secured. We tend to up the ante and spend money on more of whatever. There is never a stopping place. No matter how much we have, enough never arrives. And our thoughts do not come full circle to the fact that we lead what many would consider extravagant lives. Economic growth, however, has provided no enduring boost to human morale.

And the confusion of the more developed, advanced part of our world is only worsening the situation, for in poorer communities and entire developing countries there is an urge to improve lives by introducing the idea of working and earning more money (and spending more money), which is usually overpowering and unstoppable. We can't change this. And not only would it be pointless to try, it might be morally unsound. Who are we to say that the 800 million or so should not have their chance at a refrigerator, a telephone, a modem, or even a warm running water shower? The freedom to pursue a better life to most of us is an unalienable right.

Often, however, it is overlooked that these communities were once self-sufficient and sustainable, prior to the entry of the wheels of progress. Whole cultures are lost as the wheels of progress reshape a once working kingdom into the forever repeated urban sprawl. Many of us in the western world were taught in school that people in other cultures are less intelligent, less motivated, and less ambitious then we are. This has led to the paralyzing lack of concern our leaders have for people and animals of foreign lands, as well as to the remaining inhabitants of the U.S.[185] Thus, the ease at which we ignore past cultures and poorer communities.

The Gosinte Indians of the Utah-Nevada Great Basin were self-sufficient in their hunting and gathering ways.[186] Today's generation is now exposed to all the new fads and technologies from television, forgetting their ancestral ways of self-sufficiency and ending up as poor, government supported refugees. Look at the arrival of television toward transforming a tundra native village from its reliance on language and caribou.[187] "This was a gift from the creator.... a plane from Fairbanks was bringing the Gwich'in tribe a strange black box with aluminum antlers." Since that initial on-slough, 67 cabins in the arctic village now have TVs. Gwich'in children are so drawn by television that they do not have time to learn ancient hunting methods, their parents' language, their oral history. Instead, they dream of becoming professional basketball players, although none of them has ever touched a basketball.

And in the jungles of the Amazon, the self-sufficient, sustainable way of the traditional "rubber tapper" family is being replaced by the burning of rain forests for farming and cattle ranching.[188] Replacement of the ancestral methods have only meant hardship and poverty for once vibrant villages in the rain forests. A family of rubber tappers and nut gathers made $1,333 per year, while a family going into farming or ranching, using the same land area, made less than $800 per year. Extractive use of the forest for rubber and nuts, moreover, required no government subsidies or lender financing, which in supporting ranching and farming have caused Brazil to become a "debt peon."[189]

Many would suggest that improving the overall economic lot of a country is more important because it also beefs up the middle class, the strongest supporters of environmental responsibility. But the ever increasing Gross Domestic Product (GDP) of many developing nations leaves in its tracks a raped population whose improved education, health, and lower infant mortality have yet to be promises fulfilled. Therefore, in this approach more prosperous countries miss an opportunity for building a sustainable future. Instead we spread our addiction to consumerism, encouraging more material growth in the poorer countries, which is absurd.

Yet, any global increase in extreme material consumption, like in the country of China (one-fourth of the world's population) is ecologically unsustainable. Acknowledging this sustainability challenge is psychologically disturbing. It implies that the human race cannot safely continue on its current path, a path which by many modern values, has been stunningly successful in improving human welfare. What we are doing is advocating more growth instead of confronting the question of a fair distribution of benefits among all sectors of society. Once growth is removed as our all-purpose problem-solver, we will have to face squarely the problem of fair distribution. This is very likely to cause serious disagreements and perhaps even strife. It could get ugly—not that it isn't already when looking at the bigger picture.

Instead, pursuing goals that reflect genuine human needs, like wanting to feel connected to others, turns out to be more psychologically beneficial than spending one's life trying to impress others or to accumulate trendy clothes, fancy items, and the money to keep buying them. Sooner or later we must consider what our cravings are doing to others and will ultimately do to our children and ourselves.

*   *   *

Addicts and alcoholics go through several stages of denial, during which the consequences grow more severe always rationalizing that tomorrow is when the bad habits will be neutralized, always saying that things aren't "that bad." Addicts live for today. Until an addict lacks basic needs there continues a disregard for many of the consequences. What is always suggested to the addict in counselling is not to wait until the consequences are so large that detox is needed. It is much less stressful if we deal with these important issues while also looking down the road to what alternatives the future holds. Think about it—you make sure your child is cared for today just to be tragically killed in a school shooting or die of leukemia from drinking contaminated water a few years from now. All because you and society in general did not act in the context of care for and sustainability of future life, but instead only thought about the here and now.

MYTHS

The late and dearly missed Donella Meadows offers some interesting perspectives concerning some of the stated myths regarding why we "should" continue with business as usual.[190] In considering the following different myths, you might begin to obtain a sense for the "smoke and mirrors" that is continually being placed before society today.

Myth 1: "**Growth provides needed tax revenues**." There are a few exceptions, but the general rule is: the larger the city, the higher the taxes; the more suburban sprawl, the higher the taxes. Growth requires water, sewage treatment, road maintenance, police and fire protection, garbage pickup, and schools—a host of public services. And, almost never do the new taxes cover the new costs. New urban growth that is sprawl rarely pays its own way.

Myth 2: "**We have to grow to provide jobs**." But rarely do new jobs go to local folks. If you compare the 25 fastest growing cities in the U.S. to the 25 slowest growing, you find no significant difference in unemployment rates, because new jobs usually end up attracting more people needing those jobs, or the particular industry that grows recruits people from someplace else.

Myth 3: **"If we try to limit growth, housing prices will shoot up."** Sounds logical, but it isn't so. The important factor in housing affordability is not so much house cost as income level, so development that provides mainly low-paying retail jobs makes housing unaffordable. A study of California cities, half with strong growth controls, half with none, showed no difference in average housing prices. Some of the cities with strong growth controls had the most affordable housing, because they had active low-cost housing programs.

Myth 4: **"Environmental protection hurts the economy."** According to a Bank of America study the economies of states with high environmental standards grew consistently faster than those with weak regulations. The Institute of Southern Studies ranked all states according to 20 indicators of economic prosperity (gold) and environmental health (green) and found that they rise and fall together. Vermont ranked 3rd on the gold scale and first on the green; Louisiana ranked 50th on both.

Myth 5: **"Growth is inevitable."** There are constitutional limits to the ability of any community to put walls around itself. But dozens of municipalities (*e.g.*, in the states of Maryland and Oregon) have capped their population size or rate of growth by legal regulations, referred to as growth boundaries, based on real environmental limits and the real costs of growth to the community. Many kinds of growth cost more than the benefits they bring. So the more growth, the poorer we get. That kind of growth will kill us.

Myth 6: **"Most people don't support environmental protection."** Polls and surveys have disproved this belief for decades. In Oregon, Los Angeles, Colorado, and the U.S. as a whole for example, the fraction of respondents who say environmental quality is more important than further economic growth almost always tops 70 percent.

Myth 7: **"Vacant land is just going to waste".** Studies from all over show that open land pays far more - often twice as much - in property taxes than it costs in services. Cows don't put their kids in school; trees don't put potholes in the roads. Open land absorbs floods, recharges aquifers, cleans the air, harbors wildlife, and measurably increases the value of property nearby. We should value it and pay for it to be there.

Myth 8: "**Beauty is no basis for policy**." Dollars are not necessarily more real or important than beauty. In fact beauty can translate directly into dollars. For starters, undeveloped surroundings can add $100,000 to the price of a home.

Donella Meadows [191] goes on to say, maybe one reason these myths are proclaimed so often and loudly is that they are so obviously doubtful. The only reason to keep repeating something over and over is to keep others from thinking about it. You don't have to keep telling people that the sun rises in the east.

Short-term views propose quick fixes, resulting in superficial changes with respect to some problematic symptom, but leaving untouched the deeper cause of the symptom. Many of today's problems resulted from yesterday's solutions, and many of today's solutions are destined to become tomorrow's problems. This simply means that our quick-fix social trance blinds us because we insist on little ideas that promote fast results regardless of what happens to the system itself. People in general, have a hard time listening to those concerned with modern environmental and social issues since there seems to be an endless supply of everything, which in turn means "everything is okay." What society really needs are big, systemic ideas that both promote and safeguard social and environmental sustainability—an inside-out overhaul.

With the global nature of our society, regardless of our perspective, change is the only avenue. We are capable of acting and making beneficial changes based upon our own experiences as well as what we know. We are able to integrate change once we realize the potential influence that comprehensive correction will have.

There will always be those who embrace change for the good of all, and there will be those who are hesitant to change as they do not want to disrupt their current comfort levels. But, when we have water shortages, when we watch our children born with deformities due to the toxins in the air and food, when we realize the pain and suffering endured to make our fancy tennis shoes, and that those who inflicted that pain have no conscience, we will find the motivation to change. Once we are aware of what price is paid for our consumer demands and slanted economics, once we understand how that affects our world continually, we will set in motion a communal healing. A healing that has one direction, a destiny of achieving balance, fairness and sustainable lifestyles for all.

DO YOU FEEL ALONE?

Probably the most daunting problem facing individuals wanting to change their relationship with the natural world by reducing consumption and increasing awareness of environmental hazards, is the feeling of aloneness. Mainstream media would try to convince you that such actions are not popular.[192] The result of feeling constantly in the minority is a sense of powerlessness that stifles action, encourages compliance with perceived norms, and stimulates consumerism on a vast scale. Prepare yourself to contest the following myths promoted by those who subscribe to the mainstream beliefs of our society.

[1]   Environmental awareness is nice, but we can't afford it. The economy must come first. There isn't enough money to solve these problems. (**But the reality**…there is plenty of money in the world, just a lack of good judgement by leaders on how to use it.)

[2]   Don't worry about a few pesticides on your vegetables. It's no big deal. (**But the reality**…assaults on our bodies from environmental chemicals come from more than just vegetables…the air, the water, the soil. Any way of decreasing this assault is worth while in terms of our body's accumulation of total toxins.)

[3]   Someone or something has to suffer if others are to get what they need…afterall, it's a dog eat dog world. (**But the reality**…for the first 25,000 years that humans

inhabited Earth, life was peaceful, egalitarian, and religious-based.)

[4]   I feel fine so how can the air and water be as polluted as some claim? (**But the reality**…life can take a lot of abuse and it is often too late to "go back" after reaching a certain threshold [the boiling frog syndrome]. Why take the unnecessary risk?)

[5]   The out-of-control population growth in the developing nations is the problem. (**But the reality**…there must be a balance between consumption and population growth to have fair societies world-wide.)

[6]   Just wait…technology will solve the problem. (**But the reality**…technology takes a huge toll on natural resources and human capital that we never think about or account for.)

More material growth, better and more jobs and higher income, at least in the poor countries, seems essential for socioeconomic sustainability. And yet, any global increase in extreme material consumption, like in the country of China, with its masses of people, is ecologically unsustainable. Acknowledging this challenge is psychologically disturbing. The evidence suggests that we cannot safely continue on the current U.S. path of improving societal well-being for only the western world. There are billions of people east of here. We must advocate fairness and equal distribution of benefits for every living, breathing human being. Afterall, survival of all humans is at stake. Even the affluent need a functional, life-supporting environment.

We must continually ask ourselves what we are doing, how we are living, and what will be the ultimate outcome of our behaviors. We must question the policy analysts, decision-makers, corporate heads, stockholders, and any destructive entity if they are behaving in a way that is harmful. When testing a technology, project, program, or policy for its sustainability merits, two questions should be asked:[193]

• Will this decision or activity reduce people's ecological impact?

• Will this decision or activity improve our quality of life?

Only those decisions or activities that satisfy at least one of these criteria without violating the other can move us toward sustainability.

*   *   *

The blurring begins and ends with each individual. Facing insecurities that evolve into addictive tendencies of consumption and waste is no easy task. The imbalances, injustices, and immorality that are prevalent in the world today are not one of mental illness or physical illness or even illegalities. What we have before us is a social disease that is bore in a separatist stance, blinded by addiction, void of spiritual connection,[194] and confused by the details of science. It is said that: "No one is free until everyone is free." Looking at the state of the western world's health...mental, physical, legal, political, and environmental...there is incredible wisdom in those few words. We exist in a very addictive culture. Thus, just as the addict, we may find that simplicity is sacred.

Section IV

How Do We Act Sustainably?

Chapter 9

What is Quality of Life?

"WHAT IS SUCCESS?...To laugh often and much; to win the respect of intelligent people and the affection of children; to earn the appreciation of honest critics and endure the betrayal of false friends; to appreciate beauty; to find the best in others; to leave the world a bit better off, whether by a healthy child, a garden patch or a redeemed social condition; to know even one life has breathed easier because you have lived; that is to have succeeded."

Ralph Waldo Emerson

Albert Einstein wrote, "The world will not evolve past its current state of crisis by using the same thinking that created the situation." But, the feeling for wanting less and living more simply challenges the typical American ideal of getting ahead, of having more and bigger material things. In retrospect, however, who actually requires what consumerism societies have built into their need patterns?

- Who needs tons of throw-away packaging?
- Who needs mounds of chemical/toxic waste?
- Who needs mountains of meat?
- Who needs a free supply of drugs and arms?
- Who needs hundreds and millions of cars and a new car every year?
- Who needs unsustainable agriculture and industry?

- Who needs animal furs and exotic plants and foods?
- Who needs lifestyles built on waste and excess?
- Who needs anything that is life-threatening?
- Who needs bio-technology when the Earth's reproductive systems work just fine?
- Who needs the pollution in water, air and soil that growth usually brings?

Two hundred years ago, hot running water on demand was a luxury. Today in the West it is considered a basic need, even a "right." Is it possible that what we really should be looking for is a global view of what is meant by needs and luxuries? In the South Indian state of Kerala the incomes are very low by first world standards but well-being measures such as long life, low infant mortality, high educational levels, and low fertility are right up to the averages found in Europe. In contrast, in Kerala (as in the rest of India) fancy, designer toilet paper is a luxury found in tourist accommodations, not a need.

As never before in human history, common destiny beckons us to re-define our priorities and to seek a new beginning. It requires an inner behavioral change—a change of heart and mind. We can take decisive action to adopt, apply, and develop a collective vision for humans locally, regionally, nationally, and globally. Different cultures and communities will find their own distinctive ways to express this vision, and we will have much to learn from each other. Our best thought and action will flow from the integration of knowledge with love and compassion, for "life's a comedy to those who think, a tragedy to those who feel."[195]. Every individual, family, organization, corporation, and government has a critical role to play.

HUMAN DIMENSIONS

The first stage towards reducing devastation from our ecological impact is to realize that the main issue is about living, livelihood, and the very way of life many of us have become quite accustom to. Moving toward a sustainable future is to acknowledge that the "environmental crisis" is less an ecological and technical problem than it is a behavioral and social one. Human behavioral and attitudinal changes are the starting point to any real action to save the land, air, and water that is so vital to each of us.

Motion toward change can be characterized in many different ways. Underlying them all, however, is a common thread that recognizes we are allowing things to be lost, which should be protected for today's and future generations, and that we are failing to deal with variations in a manner which is sensitive to the needs of the environment. As we examine ways to modify our behavior, looking at the walls of emotional resistance are a beginning. Ambivalence or lack of support for action can partly be explained by general ignorance, but is caused more often by psychological and institutional barriers, conflicts with other goals, entrenched belief in alternative models, or economic dis-incentives.

But, why do we fear unwanted, uncontrollable change so much? Most commonly we resist change because we are committed to protecting our existing belief system. When we get "too comfortable" with our belief systems, we might want to think of the turtle, for which only two choices in life exist: pull its head into its shell where in safety it starves to death; or stick its neck out and risk finding something to eat and live.[196] We are most often in denial that we in fact do have a resistance to change and it is our beliefs that feed this chronic stumbling in our evolutionary progression.

Thus, the first step to getting past the mental blocks of change is to see the reality of a dying belief system. A system we can only maintain behind locked doors and closed minds. We isolate ourselves because we see change as a condition to be avoided at almost any cost. Even if it is no longer valid, it represents the safety of our past knowledge in which there are no unwelcome surprises. Change is the death of an accepted, "tried and true" belief system through which we have coped with life and which has become synonymous with our identity and therefore our security. We become defensive, fearful, and increasingly rigid in our thinking as we harden our attitudes or drink a strong shot of liquor to avoid this terrifying change needed for survival.

A second step in overcoming phobias for resisting change is to stifle our talent for developing coping mechanisms, which may mask the reality of situations. Coping mechanisms, as a strategy for survival, are often functional, positive, and entirely appropriate for a given circumstance. But they eventually can and often do become outmoded and dysfunctional as circumstances change. Clinging to dysfunctional coping mechanisms when they fail to meet current or new situations in life can lead to a hardening of attitudes, a hardening of the heart, and a rigidity that leads to destructive conflict.[197]

ETHICS and VALUES

Gandhi saw economic reasoning based on people—rather than on consumer goods—as the way to social/environmental sustainability. Whenever in doubt, apply the following test. Recall the face of the poorest and the weakest person you may have seen, and ask yourself if the step you contemplate is going to be of any use to that person? Will that person gain anything? Will it restore that person's control over his or her own life and destiny? This approach considers actions that correct things for the most disproportionately impacted people in a community.

Issues of food nutrition, health, and shelter are central to the challenge of providing an adequate standard of living for all members of the human family. These issues cannot, however, be tackled solely as technical or economic problems. Eliminating hunger and malnutrition, establishing food security, providing adequate shelter, and achieving health for all will require a shift in values, a commitment to equity, and a corresponding reorientation of policies, goals and programs.

Where poverty lives, so do many other issues.[198] There is a lack of access to safe water for 1.3 billion people. This problem only grows more severe as industries continue their unsafe dumping. Employment is a problem with 820 million people either unemployed or under-employed. In a monetary world, without a job there is not money to buy food. Inadequate food supplies affect 800 million in which 500 million are malnourished. But, people often have difficulty understanding individual threats and impacts beyond their household and community. It is easier to tune out in favor of the more immediate and manageable issues in our lives. And yet, as far as any one of the above topics is concerned, they all impact the future security and sustainability of our society—that which is preserved and maintained culturally, economically, and ecologically over the long-term.

The very things for which Western democracy is most criticized, such as loss of morality, the crisis of authority, and the reduction of life to the self-centered pursuit of immediate material gain, do not originate in democracy but in the loss of spirituality. This spirituality is the only genuine source of responsibility and self-respect. When we lose or temporarily misplace our moral and ethical standards, democracy loses much of its credibility. Unethical views of the world we live in can be characterized by the following:

      (a) wealth without work;
      (b) pleasure without conscience;
      (c) knowledge without character;
      (d) commerce without morality;
      (e) science [and technology] without humanity;
      (f) worship without sacrifice;
      (g) politics without principles;
      (h) rights without responsibility;
      (i) justice without mercy;
      (j) order without freedom;
      (k) talking without listening;
      (l) stability without change;

(m) private interests without public interests; or

(n) liberty without equality.

All the tenets of democracy are merely technical instruments, which enable people to live in dignity, relative freedom, and responsibility. They cannot in and of themselves guarantee human dignity, freedom, or responsibility. The source of these basic human potentials lie in our relationship to others and ourselves.

"Values" are persistent and collective beliefs about the way things ought to be, beliefs held both by individuals and groups of people. We all sense things differently, when we see, hear, touch, taste, and smell. Because we sense things differently, we understand them differently, and value them differently.[199] Each life, each culture, each society, however, is equally important to the evolutionary success of our world, whether we understand them or not. All differences among people, cultures, and societies are just that—differences. Notions of superiority and inferiority come from personal, familial, and societal judgments (values) of externally perceived differences.

Our values, our ethics, rather than the laws, are waking us up to the necessity for change in how each one of us live. These ethics will be the guiding light that is providing direction. Now, in this time of change, these ethics will come into play as each one of us is faced with a situation that is new. These ethics will support us in understanding the complexity of new changes. Our morality will comfort us as we struggle with changes that do not necessarily hold any instant gratification except that of doing what's right.

Ethics is essentially about how much importance we assign to ourselves, to those we know and to all that is life, whatever form or fashion…ethics that are founded in the best interests for human beings, ecosystems, and future generations. The decision for sustainable development narrows the choice of ethical reference considerably. The necessary concern for future generations and for the natural environment, upon which society depends, is no longer compatible with unrestrained selfishness or human self-centeredness. We are no longer allowing others to define what kind of people we are. We are awake and no matter how hard things get, we know at a deep, innate level there is a natural and spiritual world that should not be violated. The basic notions for achieving a safe place to live, for building an ecologically sustainable society, are happening. As we develop this society there is a transformation in our individual attitudes and values. There is now the opportunity for a complete reforming of our social structures.

CULTURAL CAPACITY

We have been asking how we might most efficiently put a chicken in every pot, a dishwasher in every kitchen, a computer in every home, and a satellite dish on every lawn. But because of our ethical reconsiderations, we are recognizing that this is not the model of development that is likely to be most beneficial in the long run.[200] It has been very difficult to get as concrete as we should about a new model of development, but we already know it extends beyond the material.

Our problems are moral, spiritual, philosophical, and behavioral. The path toward sustainable development can only be built upon a change in human psyche toward a deeper awareness. Such qualities as compassion, forbearance, trustworthiness, courage, humility, cooperation, and willingness to sacrifice for the common good, form the invisible yet now essential foundation of human society.

For society to continue being successful with the persistent decrease in all Earth's natural resources, decisions must be made that recognize and respect the Earth's carrying capacity to support its present human population, as well as the requirements and rights of future generations, and the requirements and intrinsic value of all species in nature. As we have learned in previous pages, the more materially oriented the desired lifestyle of an individual or a community, the more resources are needed to sustain it and the smaller the human population must be per unit area of landscape in order for more people to be able to live at their expectation levels. Thus, if we desire to maintain a predetermined lifestyle, we must ask new questions.

(1)   How much of any given resource is it necessary to leave intact as a biological re-investment for the health and continued productivity of the ecosystem?

(2)   How much of any given resource is necessary for us to use if we truly are to live in the lifestyle of our choice?

(3)   Do sufficient resources remain, after biological re-investment, to support our lifestyle of choice, or must we modify our lifestyle to meet what the land and water is capable of sustaining?

This view is very different from our present decisions and actions, based upon our blind faith in material progress which we think of as growth. This position, captured by the phrase "Cultural Capacity," is a conservative, other-centered concept, given finite resources and well-defined values. By first determining what we want in terms of lifestyle, we may be able to determine not only if the landscape can support our desired lifestyle but also how we must behave with respect to the environment if we are to maintain our desired lifestyle. Cultural capacity is therefore, about limits to growth of a community both in area and in absolute numbers of people.

DE-MYSTIFYING SUSTAINABLE DEVELOPMENT

The world as we know it is changing rapidly and is now at a critical turning point. There is significant uncertainty about how things will go in the next few decades. There is also growing consensus that the decisions we make as a society at this critical point will determine the course of the future for quite some time to come.[201].

Unfortunately, we as a society talk more to people across the oceans on the Internet than we talk to our neighbors. And likewise, decision-making has now become a once removed situation. Decisions are most often made with an expert's understanding of the science and technol-

ogy circumstances that presumably support the issue without appropriate interaction at the local scale where the problem might exist.

We have lost the appreciation of how delicately we are all interconnected and interdependent—from both ecological and social stand-
points. We stand in the fog in terms of being able to see and appreciate
this beauty of nature's connections. The real challenge then for people
in general, as well as governmental decision-makers, is to gain acceptance for the idea that these realities impose real constraints on the sustained growth of society.

But, in today's busy, special interest, complex world, there is a
broad range of responses to what the definition of sustainable development actually means. Uncertainties about the world around us, as
well as the contradictions many of them pose, suggest why debates
about sustainable development often are reduced so quickly into disputes about whether or not continued material expansion is feasible in
a world with limited resources. Likewise, the fact that organizations
and individuals cannot always agree on the meaning of "needs"
causes significant problems in common understanding. How can we
really know what the "needs" of future generations will be when we
can't even agree on what we ourselves "need" versus what we merely
want?

These disputes, however, can be eliminated by society acknowledging that various human uses of nature are in competition and then
showing their willingness to reconsider the makeup of economy-ecol-
ogy linkages, with an associated shift in definitions for quality of life.
Only in this way can we move beyond how the world is presently
positioned to deal with, or not deal with (deny), the problems our
global society is facing.

To succeed on our journey we must break the mystique surrounding scientific information as objective and realistic. Not to say science
is not important, because it is. Decisions now are, however, beyond
science, technology, and yes even politics. They reach beyond to also
capture our sense of relationship to others, our values, and our morality. Decisions about how we will conduct ourselves in the future, to

protect what we believe is sacred, are loaded with so many nuances. Thus, only communities most affected by the issues can understand what may or may not fit for their region and/or culture. In becoming connected again, people will benefit the most from the many different insights on sustainable actions that can come from a community and region working together.

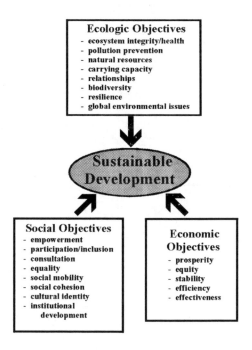

11 The interdisciplinary, cross-boundary approach to considering sustainable development requires examining all the various elements of a community's ecologic objectives, social objectives and economic objectives simultaneously.

Likewise, sustainable development requires an interdisciplinary approach in our accepting and accounting for the fact that no one specialty can provide all that is needed to address the wide range of challenges in moving toward the goal of a sustainable future. For example, the ecological, social, and economic objectives of sustainability are also concerned with the sub-issues listed in Figure 11 above. Pursuing this multi-discipline approach offers flexibility to our thinking, decisions, and actions. And as we take actions, we can learn by our mistakes and modify our actions to come closer to our defined goals.

But, achieving sustainability will also require fighting a couple of uphill battles:[202] The tragedy of the commons and the boiled frog syndrome. Most natural resources are open (in terms of access) to everyone. What is common to the greatest number gets the least amount of care and is therefore usually quickly depleted or degraded. This is the tragedy of the commons. Even if one good individual recognizes the dilemma and imminent tragedy, there is no incentive for him/her to exercise personal restraint—someone else will simply get the goods.

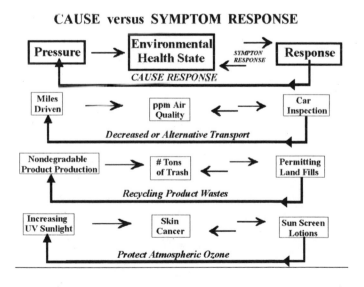

12 *A conceptual scheme that demonstrates the difference in the way society responds to the "symptoms" of a problem versus the way society should respond to the "causes" of that same problem instead.*

We are also always trying to make things simple, reducing the complexity of the world around us. In doing so, we are inclined to focus on mere symptoms of problems or on individual events, detracting us from our seeing the whole and ultimate fate and the actual cause of the problem in the first place. For example, a frog placed in a pan of water that is slowly heated will be unable to detect the gradual but deadly trend. Thus, the boiled frog syndrome. It is always much easier to fix the symptom (consequence or outcome) of a problem over the short-term than to get after the underlying cause of the problem in the first place.

But, if we tackle the causes rather than consequences (symptoms) of our local, regional, or global problems we will always come full circle to be faced with, in most instances, an environmental issue which is at the root cause of our particular problem and needing a solution for success in achieving a more sustainable community. By fixing the cause instead of the symptom, the symptoms will stop reappearing. Aspirin is not a solution for cancer, although it may make the patient more comfortable. The approach of getting to the root cause of problems rather than just treating the superficial symptoms, and the resulting outcomes, is illustrated by Figure 12.

In "developing sustainably" our strategy is to reduce our effects on nature and ecological existence, while also securing a satisfactory quality of life for everyone. In this context sustainable development should be looked at as a set of options, not a single pathway. By employing its concepts and considering its options people are aided in making informed, adaptive choices. To achieve sustainability we learn to accept our ignorance and trust our intuition, while doubting our knowledge by giving serious thought to alternative points of view.

To live safely in this world, to be sustainable, would be to:
(1)   act with the knowledge that systems (like our own body systems) have tolerances;
(2)   pay attention to early warning signs of potential overshoot of these tolerances; and

(3)   acknowledge and respect the fact that you cannot push sys-
        tems and their tolerances with abandon.

Thus, in a sustainable community, decisions about how to act
would focus on the common good that is founded on a more conservative approach...acknowledging and responding to as many warning signals as we can.

PLAIN LIVING: LOCALIZE RATHER THAN GLOBALIZATION

Above all else, sustainability requires that we reclaim the economy
in the service of people and their communities. In contrast, globaliza-
tion is creating a world of "powerless places at the mercy of placeless
powers." The purpose of economic activity should be to foster material security where people live rather than to promote out-of-sight,
out-of-mind consumption to maintain the world's financial centers at
the expense of the ecosphere.[203]

It may seem paradoxical, but global security is likely to find its
deepest roots in strengthened local community and regional
economies. No power on Earth can manage globally. Bio-region is a
critical concept in evaluating our ecological impacts elsewhere if communities are going to practice the economics of social/environmental
sustainability. Without a collective vision of sustainability within a
well-defined bio-region, the communities are no more than economic
colonies for the national and international economic markets. If individual bio-regions learn to live on the sustainable use of their own
resources, supplemented by ecologically balanced trade, the net effect
would be global sustainability, which can offer a good quality of life
for everyone. Anything short of this will one day lead to global catas-
trophe.

As communities develop industries that are scaled for their local
community and learn to supply a majority of their own needs, these
communities can then become more than a colony of the national or

global economy. In rural areas, communities must develop small-scale industries and businesses to support and add value to the local farm and/or forest economy. The community must endeavor to produce as much of its own energy as possible. The community must strive to increase earnings (in whatever form) within the community, and decrease expenditures outside the community. Money paid into the local economy must circulate within the community for as long as possible before it is paid out. Likewise, a sustainable rural community will be dependent on the urban consumers' loyalty to local products. We are thus talking about an economy that will always be more cooperative than competitive.

Remember our discussion earlier of comparison between the energy from the sun entering the desert ecosystem and that same energy entering a forest ecosystem? The desert ecosystem quickly passes the sun's energy through, using very little, and thus is not very productive. The forest ecosystem, on the other hand, recycles the sun's energy as many times as there are different and diverse species assemblages to use it, before the energy leaves the system. This recycling and reuse make the forest ecosystem much more stable and productive.

Likewise, sustainable human community development increases adaptability and productivity by creating and maintaining a diversified social and economic base with locally shared ownership and access to basic, co-developed human services. Community adaptability, and therefore stability, is based on its ability to meet the majority of its own needs within itself, instead of being dependent on outside sources, through the diversified co-development of social and economic institutions. This means, however, that the adaptability of a community also encompasses the ecological integrity of its surrounding landscape. This local control is required to:

- manage human-ecosystem relationships in a manner that is sensitive to local conditions at the scale of ecosystems,
- re-establish a sense of "connnectedness" between human populations and the ecosystems that support them, and

• reduce the alienation that intrudes between people and their employment when resources and capital are owned or controlled by absentee landlords.

Forcing the issue requires we the individual learning to curtail our consumption at a minimum. At it's fullest potential, we the individual must take control back of our own life. We can gain this control by means of supporting ourselves through local community. People can also stretch their limited ecological economic resources even further by practicing conscious simplicity. The idea of wanting less, and living more simply on what local resources can provide, however, challenges the typical American ideal of getting ahead, of having more and bigger material things.

Consider this Indonesia Story…despite enormous momentum to the adoption of outwardly western life-styles around the world, there has been a deep ambivalence in places like Indonesia. Many of the more thoughtful Indonesians, especially those associated with Islam, feel that the adoption of Western consumerism brings with it a corruption of some of the traditional values—of family closeness, respect for elders, communal solidarity, and moral purity. The point is to ask that we do not take refuge in technological solutions, but seriously debate prospects for social transformation that give much greater recognition to the hollowness of global consumerism and the duty to care for the natural environment.

*   *   *

Living more simply does not mean deprivation or hardship. It means focusing on what is **sufficient** for comfort, hygiene, efficiency, etc.[204] Most of our basic needs can be met by quite simple and resource-cheap devices and ways, compared with those taken for granted and idolized in a global-wide consumer society. Living in ways that minimize resource use should not be seen as an irksome effort that must be made in order to save the planet. These ways can become important sources of life satisfaction. We can come to see as enjoyable many activities such as living frugally, recycling, growing food, making rather than buying, composting, repairing, giving old things to others, making things last, and running a relatively self-sufficient household economy.

A community that embraces plain living and therefore sustainability, is a place where we grow our own vegetables and support the local farmers to supply us with our meats. Likewise, it is a place where we know who made the clothes we wear and contribute in a manner in which we see results within the space in which we live.

Now is the time to practice conscious living—doing the right thing while we have choices. A globalized economic system has an inherent bias in favor of the large, the global, the competitive, the resource extractive, and the short-term.[205] Our challenge is to create a global system that is biased toward the small, local, cooperative, resource-conserving, and long-term...one that empowers people to create a good living in balance with nature.

Chapter 10

Science, People, and Community

"Cherish your vision: Cherish your ideals: Cherish the music that stirs in your heart, the beauty in your mind, the loveliness that drapes your purest thoughts. If you remain true to them, your world will at last be built."

James Allen

A piecemeal fragmentation has occurred in people's lives and the way they view the world around them. The influence of this fragmentation on our thinking disintegrates our social structure by obliterating the sense of community as a living system. Starting with the agricultural revolution, and continuing through the industrial revolution, increasing separation in the social order has produced a progressive fragmentation in belief systems and thought. And yet, all the people in a community group come from different experiences and these experiences combined together can collectively assist the whole group in more awareness and understanding. That is to say that the effective function of the group defines the arrangement, which in turn defines the structure, and it is by the visible structure that we tend to characterize a system, and the many strands that comprise the web of life for that system.

UNDERSTANDING THE VOICE OF SCIENCE

As a global society, those of us in a position to begin doing something about pending environmental problems are distracted, more so than ever before, by the constant barrage of information on potential prosperity. We are being lulled into a false sense of power and security by the offerings of the Internet. It would be a real mistake to confuse the vibrancy of the virtual world with the increasingly troubled state of the real world.[206] Science can certainly help us here. But, proceed with caution.

Science is the most exciting and sustained enterprise of discovery in the history of the human species. It is the great adventure of our time. In a stunningly short period of time, science has extended our knowledge all the way from the behavior of galaxies to the behavior of particles in the subatomic world.[207] Science is an indispensable expression of human intelligence and will, and is appreciated as a systematic investigative approach to developing knowledge.

But science can not function in isolation. Science and moral beliefs should share a common premise. Science tries to document the factual character of the natural world, developing theories that coordinate and explain revealed facts. Moral beliefs, on the other hand, operate in the equally important, but different, realm of human purposes, meaning, and values...subjects the factual domain of science might illuminate, but can never resolve. Therefore, science and ethics should coexist peacefully and respectfully.[208]

It is through the balanced combination and cooperation of science and morality that humanity can acquire a genuine humility and respect for nature, while applying the appropriate skills and technologies needed to advance civilization. Can we move beyond notions of correlation, to a more sophisticated appreciation of science and ethics as representing complementary ways of understanding our universe...both Earth and the human condition? Only in this way of combining knowledge of both our spiritual growth and our common dependence on the ecosphere, are we called to be, fully and consciously, citizens of one Earth home.

The shift that needs to occur is one of bringing awareness and understanding to the people who can make the most difference, the average global citizen. Unfortunately, ecologists often lack the philosophical, theoretical, and empirical foundations for the many issues facing our global environment. Even when ecologists do know enough to converse and suggest social actions to alleviate problems, they are often dreadful communicators, unaware of people and politics, and lack talent for the simple communication of complex ideas.[209]

Likewise, in dealing with sustainable solutions to problems faced by a community of people, it is unbelievable how much the people that live and work in that community really know. Without many people looking at a problem and bringing their different viewpoints to bear on it, errors remain uncorrected, narrow perspectives and selfish motives are rewarded, and the general welfare will not usually be promoted. For that to happen, the concept of sustainable development must be articulated in terms familiar to community/business leaders as well as individual citizens. The challenge lies in looking for solutions about how institutions could work more efficiently, more collaboratively, more respectful of the public way of gaining knowledge in any particular place or culture. Instead of there being them and us, let's get the job done in a way that is collaborative.

An important avenue to implementing sustainability also comes from the conduct of research by and on behalf of the community. This can involve what has now become know as "citizen science,"[210] which aims to engage the expert way of knowing (scientists) with the public way of knowing (stakeholders) in an ongoing dialogue with environmental managers about the kinds of ecosystems people want and the kinds of ecosystems people can in reality get. Community-based research processes involving citizen science, unlike mainstream scientific and technical research, are very valuable to community groups that are eager to know the research results and to use them in practical efforts to achieve constructive social change.

A good example of this process comes from looking at an event that took place in Africa over the last 10 years. An African village began cultivating timber bamboo for building construction. They planted this in places where the rich soil needed for a farm had been eroding away. By harvesting the bamboo at maturity, which was planted three years earlier, the community constructed its buildings and shelter, using a material that is readily renewable. The community has grown economically and materially without any deleterious effect to the environment. Planting timber bamboo to prevent soil erosion, while using this renewable resource for building materials, was a perfect community-based research approach to achieving sustainable development.

It is within the context of this iterative process of social change, allied with the application of scientific knowledge, guided by moral, spiritual principles, that truly sustainable models of production and habitation can emerge. In other words, the advancement of a new set of values for nature and the unfolding of a global order that can effectively address environment and development issues, both depend on the enhancement of the only infinite resource in the face of depleting material resources—human spiritual potential.

Development within each individual of spiritual qualities such as respect, compassion, selflessness, creativity, and motivation to serve is essential to building a global society that can maintain higher levels of unity and motivate the fundamental changes in structures and values required in order to follow an ethic of sustainability. It is with the application of scientific knowledge, guided by moral principles, that truly sustainable models of manufacturing, consumerism, and human habitation of nature can emerge.

REFRAMING THE ISSUE

The use of science should also be shrouded in an environment of adaptability, for the most well intentioned conservation activities can run afoul of what is really happening in nature. Take for example the categorical approach to forest fire suppression.[211] A no-burn policy may increase a forest's fuel load to the point at which a lightning strike produces a huge, uncontrollable fire that not only wipes out thousands of acres of forest but also takes a lot of homes with it as well. That's outright failure: a catastrophic fire consumes stands that survived centuries of natural, periodic, small scale fire cycles before no-burn policies were prescribed. Without burning, not only does fuel for a more devastating fire build up, but fire-tolerant, desired species in a forest will eventually also begin to disappear, as they are replaced by species better adapted to the absence of fires...part of nature's story of evolution.

The key to controlling these demons of "humans can fix it" strategies that science will often offer a false sense of security for, might be to do a better job of managing systems in their entirety. Management without understanding leads to problems.

An illustrative example would be the spruce tree budworm outbreak in the 1970s.[212] Successful suppression of spruce budworm populations in eastern Canada, using insecticides, certainly preserved the pulp and paper industry, and employment, in the short-term by partially protecting the stricken forests. But, the elimination of short-term, isolated disease outbreaks that natural processes might be primed to respond to left the door open a few years later for a more devastating outbreak over a much larger area and of an intensity never experienced before, resulting in massive economic losses. More recent examples include livestock problems in Western Europe, Mad Cow Disease, the pig plague, and poisonous chickens in Asia.[213]

In contrast, if we had understood the spruce tree budworm ecology in the 1970s, as we do today,[214] ecosystem management would possibly have been conducted differently, with a different economic result. When budworm outbreaks first occur in boreal forests, small scale, localized bird populations, such as warblers, respond by increasing their numbers because of the food availability that budworms represent. When local warbler feeding is unable to control the budworms and their densities increase and aggregate, larger birds, such as thrushes, begin to prey upon them. As the outbreak progresses further and entire stands of trees are involved, the largest birds in the affected region begin to eat budworms, as these budworms now represent an aggregated large-scale perturbation. In considering the full evolution of a particular budworm outbreak, these different sizes and functions of birds in the habitat combine to provide a strong and highly resilient control of budworm populations over the long-term.

An analogy to the budworm scenario even occurs with regards to the human immune system. A localized infection in the human body first causes a local response from a few immune cells. If that response is unsuccessful, a broader response occurs, and the immune system begins to manufacture more immune cells. As a pathogen invasion progresses in the body, other systems become involved and shivering may begin in the infected person, as a **whole system** response to the pathogen invasion, similar to the whole system response to budworm increases by a diversity of bird predators.[215]

There is also the story of the Florida Everglades. In the past referred to as a "River of Grass," the Everglades was once a huge ecosystem in south Florida (USA) that included marsh grasses, cypress trees, herons and other birds, fish and small aquatic animals, alligators, and panthers. Agriculture and mushrooming urban centers ate away half of the original Everglades extent. But on top of this, after a devastating flood in 1948 the U.S. Army Corps of Engineers built a system of levees and canals to curb flooding and create a dependable water supply to support further human development. As a result the Everglades has become drier and its many kingdoms are suffering. And if that is not enough, the very productive ocean waters off the coast of Florida have been damaged by too much fresh water flow to these salty systems from the many changes in the system's hydrology.[216] Through this process fresh water has become a "pollutant." Now the Corps of Engineers hopes to undo its mistake…at an estimated cost of $11 billion to taxpayers.

PEOPLE CONNECTIONS

We believe we are omnipotent and don't need others, so we manage to ignore terrorism, climatological change, collapsing governments, chemical spills, plagues, recession and failing banks, floating debris, the disintegrating ozone layer. Volcanoes, earthquakes and hurricanes, religious feuds, defective vehicles and scientific charlatans, mass murderers and serial killers, tidal waves of cancer, AIDS, deforestation and exploding aircraft are as far away from our remote lives as braid catches, canions, and rosette-embroidered garters.

With increased specialization, however, people have become more interdependent, while at the same time being less able to deal with the complexity of relationships in their lives individually. While we have become more complexly interrelated on a global scale, we have also become more "distant" from each other and our environmental problems on a local or regional scale. From our increasingly specialized understandings and particular points of view, often developed in isolation from others, it is difficult to comprehend the significance of working together to govern resources that we all share in common. And given these complexities, finding fair solutions for the large problems that confront all of us is even more challenging. The answer lies in recognizing the value of relationships.

Relationship is the essence of humanity because we are creatures who must share to find value, regardless of how it is defined. Today there's a real emptiness, a real aloneness in people's lives. We were never meant to be alone, and the basic desire of people is to be bonded with others and to have meaningful relationships. **"If you want a sustainable future for communities, you need a sustainable family."** [217]

The truth is, existence for all life is encircled in its interdependent, interactive relationship to everything else. We must start to understand that change from a consumer oriented thought process to a systems approach is required. The quality indicators of both ecological and economic health are rooted in the quality of the relationships between and among the parts, as well as how these relationships change and evolve towards things working better, whether it be a natural ecosystem or a community of people. As such, every relationship is dynamic, constantly adjusting itself to fit precisely into all other relationships, which consequently are adjusting themselves to fit into every other relationship, and so on into infinity.

The heart of sustainable development also involves joint ownership of the process for each person involved. In becoming aware of the many sides of science and beginning to understand our impacts on nature, no one can empower anyone else—one can only empower oneself. One can, however, give others the psychological space, permission, and skills necessary to empower themselves and then support their empowerment and taking of ownership in a situation. What a feeling of good and responsible citizenship this can cause.

Central to the notion of empowerment is people's willingness to accept responsibility for their own behavior, **first** by overcoming internal barriers to change, **second** learning to work with one another for a greater good, and **third** by directing formal authority within the democratic process.[218] Thus, if it weren't for all those folks with badgering ways and annoying cries from the sidelines, the great clunky machinery of status quo would never move at all. It would grind away as it always has, day after day, slowly replicating itself, slowly reproducing today the exact same results it produced yesterday.[219]

We are not only responsible for ourselves, but for our families, extended families, and ultimately each other. Neighborhood watch systems are successful outcomes of people collectively calling from the sidelines. In times of hardship we see this coming together of people. And ownership demands a commitment to be as informed as possible about the whole. Because owners cannot walk away from their concerns, everyone's accountability begins to change.

Because sustainable development honors the integrity of both society's experience and its environment, the outcome of people interacting with people is a unified world view in which a system's function defines the system. Experiential transfer among people is thus critical to understanding how ecosystems and their interconnected, interactive components function, including the bridge between a community and its surrounding environment. And then we discover another way, one that increasing numbers of people are considering. It is an avenue of "enough", a pathway for the good life determined by what we do and how we relate to others, not by how much money we make or how much stuff we own.[220]

DEEPENING PEOPLE'S SENSE OF COMMUNITY

The means to effect change in communities has traditionally pitted advocacy groups, elected leaders, and influential citizens against one another. Leaders usually focus on bringing together groups of like-minded people or special interest coalitions to overpower others. When this strategy works it leaves people divided. When it does not work it leaves gridlock. Both function to alienate people from the process of further involvement.

The concept of civic duty, however, is not dead. People still want to participate; they just can't figure out how to. Everyone has a piece of the truth in striving for solutions that break out of the tired legacies of compromise, Robert's Rules, winners and losers, and the politics of conflict, where every issue is either right or wrong. We must understand the benefits that come from laboring over consensus, where consensus means that all parties, no matter their stature or power in the conventional sense, can be confident that their view of things, their piece of truth, will be heard. True attempts at building consensus resist at all cost the thought that there are people who have nothing of significance to offer.

Can we learn to listen to one another's ideas, not as points of debate but as different and valid experiences in a collective reality? Although one may not think of it as such, listening is the other half of communication. Communication is a gift of ideas; therefore, the other person can give me a gift of ideas through speaking only if I accept the gift through listening. If we listen to one another and validate one another's feelings, even if we don't agree, differences will be resolved before they become disputes.

Unyielding self-centeredness represents a narrowness of thinking that prevents cooperation, coordination, creative thinking, and the resolution of issues. Unity on the other hand, is the instrument whereby true justice can be established, whereby equality of opportunity and privilege can exist for women and men, as well as for different races or for different ages, throughout the planet. For sustainability to be possible therefore, self-centeredness must blend into other-centered teamwork. This involves setting aside egos and accepting points of view as negotiable differences while striving for the common good over the long-term, the ingredients necessary for teamwork and a sense of unity.

A good example of unity is the following. When asked about their ethnic backgrounds, Malaysians living in a great mix of cultures, national origins, and religions in a very small space, answered with a response that reflected their national unity. They referred to themselves as Malaysian Chinese, Malaysian Indians, Malaysian Sri Lankans, and so on. When a person in the U.S. is asked this question, however, their response is usually that they are an Afro-American, Chinese-American, Japanese-American, German-American, Italian-American, and so on. While the difference may be subtle, it is profound. The Malaysians focus on their unity, while we in the U.S. focus on our sense of separation.

But, when everyone's future is at stake, community pops into place: historic alliances and differences among people suddenly fade. The Chinese adage says that the word for crisis in that language has two symbols or characters: one for danger, the other for hidden opportunity. Crises shake people enough to let them break free of old conceptions. Under these circumstances we can then respect the integrity of different cultures and really look hard for the opportunities that present themselves, first by rejecting the idea that nature is merely a collection of resources to be used.

Then, we can realize that our social, economic, environmental, and spiritual problems are interconnected, causing us to cooperate in developing integrated strategies to address them. Through community-wide actions we can resolve to balance and harmonize individual interests with the common good, freedom with responsibility, diversity with unity, short term objectives with long term goals, and economic progress with the continual flourishing of ecological systems.

The courage and the willingness of many people in a community to adopt and implement the behaviors that lead to a sustainable lifestyle are the beginning. This move will lead to a social evolution—an evolution in which change is accepted as a process to be embraced. As understanding of sustainability increases, the notion of ever-adjusting relationships becomes a creative energy that guides a vibrant, adaptable, ever-renewing society through the present toward the future. The potential ability to transfer results of experiences from here to there, and from one person to another, is influenced by the breadth of one's experiences and catalyzed by relationships. The result is shared solutions to shared problems that could not become obvious without working together.

During any community consultation process, a distinctive method of non-adversarial decision-making, the use of science has its place and should reign accordingly, not overwhelming the citizen's own way of knowing. Computer technology and quantitative modeling can help provide solutions to natural resource and ecosystem management problems. Western trained scientists, however, often do not appreciate the extent to which solutions depend on the expertise and power of local people.[221] Native knowledge provides direction for data collection, villagers' priorities guide the formulation of environmental management questions, and community institutions implement policies. Science provides the tools and the town or neighborhood people provide the public way of knowing, the incentive, and the momentum.

Likewise, community unity calls for a reflection by people on the fundamental oneness in the whole of creation and requires a new understanding of the relationship of parts to each other and to the whole. Notions of superiority and inferiority are based on personal, familial, and societal judgments about the intrinsic values of extrinsic differences. If we do not learn to focus on our common aspirations, our unity, we will continue to focus on our differences, which translates into all kinds of segregation—not on paper perhaps, but in the streets, where racism for example, cannot hide. Sustainable development insists upon promoting feelings of self-worth in people (equality with others), allowing them to act as catalysts in the process of change without intimidation, whether in their own lives or in the life of society. For change to be a creative process, each person must respect every other person, in a sense of unity, as well as the intrinsic value of his or her environment.

WHAT PAIR OF GLASSES ARE YOU LOOKING THROUGH?

An intrapersonal relationship is the relationship that exists within ourself. It is our sense of spirituality, self-worth, personal growth, and so on. In short, it is what makes us conscious of and accountable for our own behavior and its consequences. It basically influences the way we think about things. When we become spiritually conscious the more other-centered we become, the more self-controlled our behavior is, and the greater our willingness to be personally accountable for the outcome of one's behavior with respect to the welfare of fellow citizens, present and future, and the Earth as a whole.[222]

Self-centered people tend to be piece thinkers. Other-centered people tend to be systems thinkers. A systems thinker is likely to see himself or herself as an inseparable part of the system, whereas a piece thinker normally sets himself or herself apart from and above the system. Like blind people feeling the different parts of an elephant, each person is initially limited by his or her own perspective. For example, a glass of water is half full or half empty, depending on one's point of view, but the level of water is the same in either case. Likewise, a beginner, unfettered by rules of having to be something special, sees only what the answers might be and knows not what they should be.[223] Understanding that our imagination is bounded by blind acceptance of social convention frees a willingness to reach beyond such convention, opening the soul's creative eye. Most social-oriented individuals become stuck within self-imposed limitations, concerned most about what people around them might think. Unfortunately, it is the rare individual who has managed to retain a childlike beginner's mind.

Table 1. Differences between the indigenous Native world view and the western world view.

Native	Western
Spirituality is imbedded in all elements of the cosmos.	Spirituality is centered in a single Supreme Being.
Humans have responsibility for maintaining harmonious relationship with the natural world.	Humans exercise dominion over nature to use it for personal and economic gain.
The universe is made up of dynamic, ever-changing natural forces.	Universe is made up of an array of static physical objects.
Time is circular with random natural cycles that sustain all life.	Time is linear chronology of "human progress".
Nature is honored routinely through daily spiritual practice.	Spiritual practices are intermittent and set apart from daily life.
Respect for elders is based on their compassion and reconciliation of outer- and inner-directed knowledge.	Respect for others is based on material achievement and chronological old age.
Sense of empathy and kinship with all forms of life.	Sense of separateness from and superiority over other forms of life.

Consider the possibility that there are **at least** two different ways of seeing the world. On the one hand there is the self-centered, western view. And on the other, there is the more socially sensitive, innocent native world view. Look at the comparison of these two views in the Table above and decide which road you would prefer.

Which view do you prefer? It all depends upon what kind of eye glasses you are seeing the world through—what kind of experiences support and reinforce your viewpoints. These experiences might be different from your neighbor's because you see things through different glasses. Does this make one of you wrong? Not really. But it does open the door for possible dialogue between you and your neighbor concerning the realm of possibilities.

Being fully conscious in the present is also important, because fear is a projection of a past experience into the future. One cannot, therefore, be afraid in the present, in the here and now, because the present moment cannot be projected into the future. Only the past can be projected into the future. What one really fears is the future, the fear of the unknown, the fear of being afraid. Struggling to keep yourself in the present is thus a conscious choice one must make each time fear raises its ugly head.

Another reason for being in the present is to allow someone else to be in the present with you. If one person is in their past and another in his/her past, you are probably not in the same place or time. Such presence in the moment is a prerequisite for dual cooperation and mutual understanding for the process of growth in awareness. What pair of glasses are you seeing the world through today? What happens if you exchange your glasses with your neighbor? Do you see things differently then with your own glasses on?

TRANSFORMING COMMUNITY INTERACTIONS

The loss of community interaction is indeed a primary barrier to sustainability. Devoting far more attention to local and regional economies and politics will be the beginning. Rebuilding people relationship networks will increase understanding. In the search for sustainable economics, what is needed is a "bottom-up" society, a community of communities that are local and relatively small. The idea is to detach communities from the market economy and rebuild political responses household by household, neighborhood by neighborhood, community by community. Communities grow when individuals "commune", "communicate", and earnestly search for "commons".

Before beginning projects one tends to look over the project as a whole. What will be needed, the time it will take, the goals of the project, and the benefits of those goals. The same would hopefully stand true when considering the affects of our actions on nature. Before altering the environment, on whatever scale, one would hope that the consequences of a decision are fully considered and taken into account. **Sustainability stands strongly on the premise of forethought.** Sustainability requires that a community try to understand the behavior of the natural systems that sustain it...which makes total sense. Also, sustainability accepts that its' knowledge will always be incomplete. The endurance of sustainable philosophy will hopefully take people away from a view of the Earth and its resources that these exist purely for humans, and instead bring us to a belief that humans are part of the living Earth.

Always ask of any proposed change or innovation: what will this do to our community, present and future? How will this effect our common wealth, present and future? The sustainable community must rely on its most informed understanding of a situation, its commitment and love of home place, and long-term economic interests to establish workable limits to resource use. Establishing limits and understanding the effectiveness of these limits would then become the true practice of sustainable resource management. A community in balance with its natural resource base also understands that change is both inevitable and unpredictable. Thus, communities practicing sustainable resource management and acting cautiously would not have such rigid and fixed expectations of their environmental resources that they would have no ability to be flexible in the face of change.

The "foundation blocks" we define to live by are critical to the conduct of a sustainable lifestyle. The underpinning elements for sustainability can all be found in the definition of an "Earth Charter" prepared in 1999 by the Earth Council.[224] Excerpts of this Charter are reprinted here in order to provide guidance on the comprehensive, integrated nature of our required thinking in order to begin envisioning a more sustainable lifestyle for each individual to enact.

*   *   *

THE EARTH CHARTER

In our diverse, yet increasingly interdependent world, it is important that we the people of Earth declare our responsibility to one another, to the greater community of life, and to future generations. We are one human family and one Earth community with a common destiny, no matter where we call home. The Earth community stands at a defining moment. With science and technology have come great benefits and also great harm. Fundamental changes in our attitudes, values, and ways of living can be implemented with the right guidance.

We **can** respect the integrity of different cultures. We **can** treat Earth with respect, rejecting the thought that nature is merely a collection of resources to be used. We **can** realize that our social, economic, environmental, and spiritual problems are interconnected and cooperate in developing integrated strategies to address them. We **can** resolve to balance and harmonize individual interests with the common good, freedom with responsibility, diversity with unity, short-term objectives with long-term goals, and economic progress with flourishing of ecological systems. To fulfill these aspirations, we must recognize that human development is not just about having more, but also about being more. The challenges humanity faces can only be met if people everywhere acquire an awareness of global interdependence, identify themselves with the larger world, and decide to live with a sense of universal responsibility. The choice is ours...to care for Earth and one another or to participate in the destruction of ourselves and the diversity of life.

General Principles:

Respect Earth and all life. Recognize the interdependence and intrinsic value of all beings. Affirm respect for the inherent dignity of every person and faith in the intellectual, ethical, and spiritual potential of humanity.

Care for the community of life in all its diversity. Accept that responsibility for Earth is shared by everyone. Affirm that this common responsibility takes different forms for different individuals, groups, and nations, depending on their contribution to existing problems and the resources at hand.

Strive to build free, just, participatory, sustainable, and peaceful societies. Affirm that with freedom, knowledge, and power goes responsibility and the need for moral self-restraint. Recognize that a decent standard of living for all and the quality of relations among people and with nature are the true measure of progress.

Secure Earth's abundance and beauty for present and future generations. Each generation must conserve, improve, and expand their natural and cultural heritage and transmit it safely to future generations.

Ecological Integrity:

Protect and restore the integrity of Earth's ecological systems, with special concern for biological diversity and the natural processes that sustain and renew life. Make ecological conservation an integral part of all development planning and implementation. Establish nature and biosphere reserves to maintain Earth's biological diversity and life-support systems. Sustainably manage the extraction of renewable resources such as food, water, and wood to secure the resilience and productivity of ecological systems. Promote the recovery of endangered species and populations and protect against the human-mediated introduction of alien species into the environment.

Prevent harm to the environment as the best method of ecological protection and, when knowledge is limited, take the path of caution. Consider the cumulative, long-term, and global consequences of individual and local actions. Establish environmental protection standards and monitoring systems with the power to detect significant human environmental impacts. Mandate that the polluter must bear the full cost of pollution. Ensure that measures taken to prevent or control natural disasters, infestations, and diseases are directed to the relevant causes and avoid harmful side effects. Take all reasonable precautionary measures to prevent transboundary environmental harm.

Treat all living beings with compassion, and protect them from cruelty and wanton destruction.

A Just and Sustainable Economic Order:

Adopt patterns of consumption, production, and reproduction that respect and safeguard Earth's regenerative capacities, human rights, and community well-being. Eliminate harmful waste, and work to ensure that all waste can be either consumed by biological systems or used over the long-term in industrial and technological systems. Act with restraint and efficiency when using energy and other resources, and reduce, reuse, and recycle materials while constantly increasing the use of renewable energy sources such as the sun, the wind, biomass, and hydrogen. Establish market prices and economic indicators that reflect the full environmental and social costs of human activities, taking into account the economic value of the services provided by ecological systems. Provide universal access to health care that fosters reproductive health and responsible reproduction.

Ensure that economic activities support and promote human development in an equitable and sustainable manner, eradicating poverty, as an ethical, social, economic, and ecological imperative. Promote the equitable distribution of wealth. Assist all communities and nations in developing the intellectual, financial, and technical resources to meet their basic needs, protect the environment, and improve the quality of life. Establish fair and just access to land, natural resources, and credit, empowering every person to attain a secure and sustainable livelihood. Recognize the ignored, protect the vulnerable, serve those who suffer, and respect their right to develop their capacities and to pursue their aspirations. Relieve developing nations of onerous international debts that impede their progress in meeting basic human needs through sustainable development.

Honor and defend the right of all persons, without discrimination, to an environment supportive of their dignity, bodily health, and spiritual well-being. Secure the human right to potable water, clean air, uncontaminated soil, food security, and safe sanitation in urban, rural, and remote environments. Establish racial, religious, ethnic, and socioeconomic equality. Affirm the right of indigenous peoples to their spirituality, knowledge, lands and resources and to their related practice of traditional sustainable livelihoods. Support scientific research in the public interest. Assess and regulate emerging technologies, such as biotechnology, regarding their environmental, health, and socioeconomic impacts and guarantee their access to all peoples.

Democracy and Peace:

Establish access to information, inclusive participation in decision making, and transparency, truthfulness, and accountability in governance. Secure the right of all persons to be informed about ecological, economic, and social developments that affect the quality of their lives, ensuring these remain accessible and in the public domain. Enable local communities to care for their own environments, and assign responsibilities for environmental protection to the levels of government where they can be carried out most effectively.

Affirm and promote gender equality as a prerequisite to sustainable development. Make the knowledge, values, and skills needed to build just and sustainable communities an integral part of formal education and lifelong learning for all. Encourage the contribution of the artistic imagination and the humanities as well as the sciences in environmental education and sustainable development.

Create a culture of peace and cooperation. Practice nonviolence, implement comprehensive strategies to prevent violent conflict, and use collaborative problem solving to manage and resolve conflict. Teach tolerance and forgiveness, and promote cross cultural and interreligious dialogue and collaboration. Recognize that peace is the wholeness created by balanced and harmonious relationships with oneself, other persons, other cultures, other life, Earth, and the larger whole of which all of us are a part.

*   *   *

We will succeed through a new awakening to reverence for life, a firm commitment to restoration of Earth's ecological integrity, a quickening of the struggle for justice and empowerment of all people, cooperative engagement of global problems, peaceful management of change, and joyful celebration of life. It is most important to note that the ideas expressed above are not just about environmental protection, but cut across the gambit of our global society's life issues. All these elements of a proposed Earth Charter must come together to achieve a more sustainable and healthy world for all peoples and all other life forms, now and in the future.

Section V

Sustainable Development in Action

Chapter 11

Where Do We Start?

You cannot control the things that happen to you in life. It is rather about how you handle the things that happen to you in life. Likewise, it is not where you live, but rather how you live.

One cannot find a place at the present time where sustainability is perfectly practiced. What you will find is the budding beginnings of sustainability. Likewise, there is not a perfect place to practice sustainability as everywhere is perfect in its own way. Every area has its niche. Thus, every attempt and success is uniquely different and innately perfect. And, it is growing as the industrial revolution once did. For sustainability is like pure love and equality: grand goals that should always lead human endeavor, higher thoughts, and conscientious behavior.

Likewise, sustainability is not a destination at which we will arrive; it is a process, a mystery, and an adventure. There will be obstacles, but these should not derail us, for our journey toward sustainability, like the journey toward human equality, is forever evolving. Sustainable development is not something that is legislated, but rather something that becomes a chosen lifestyle at an individual citizen level. Through this choice we the individuals of our countries, we the working class, will have our constitutional freedoms back...the revolution of the new Millennium.

TAKE A FIRST STEP

We start with baby steps. As individuals, as neighborhoods, as communities, counties, states, and as nations we begin to take baby steps. Small changes scattered here and there. That one person who will take the reins of a particular area and lead. This is the new beginning. As leaders or as followers, we are open to change. We are open to new beginnings.

Beginnings toward sustainability such as that roof top garden in Detroit Michigan. Beginnings of sustainability such as the people of other sky-scrapers taking the same initiative, no longer consuming imported fruits and vegetables, having no idea where they have been, who they were grown by, or what might have been sprayed on them. Beginnings of sustainability such as the school yard gardens, where the children are taught how to grow their own food for school lunches.[225] And, where it is canned and preserved on site as children learn the process of balance and planning for long cold winters. Where the children themselves put their hands in the soil and gain so much more than the food that they will harvest.[226] It is these very schools, taking the initiative to make sure children are eating healthy organic, non-pesticide laced food, that are the hero's.

Beginnings are being born everyday, where change for the better brings new light and growth. Sustainability is set in motion from the bottom up, from the specifics of small projects to the view of the big picture, from our homes, to our neighbors, to across oceans. Baby steps that incite change globally. Motion that will begin to heal the world.

The healing is set in motion with the revival of the family farm, getting in touch with the land once again. Sustainability blossoms from those farmers who refuse to use pesticides. Sustainability is every farmer's market on the globe. It is Europe's position not to buy genetically altered soybeans and corn. It is that lone windmill by the bay that is pumping energy for an entire neighborhood, or the wind farm by the ocean that is creating energy for the county. Sustainability is that playground in Gaviotos Columbia that pumps water for the school from the seesaw as the children bounce up and down.[227] It is the elementary school in Yellow Springs Ohio that is heated and cooled by the solar panels that sit near the front entrance. It is an electric car.

It is that one house in a suburban sprawl that is using a solar heated water system. It is the noble warrior who built her off-the-grid house facing south, not knowing that there is such an animal as true north, who now has crooked solar panels proudly mounted...though a bit off kilter...on the roof. She lives that harmless existence.[228] New examples are being born every day. Change is occurring; baby steps are turning into 20 mile marathons.

And then there are the fields of windy dreams.[229] A farmer in Iowa who leases a quarter acre of cropland to the local utility as a site for wind turbines can typically earn $2,000 a year in royalties from the electricity produced. In addition, since the wind turbines scattered across the farm or ranch do not interfere with the use of the land for many forms of farming or cattle grazing, in a good year the same quarter acre plot can produce $100 worth of corn.

U.S. wind generating capacity for electricity production has increased 29% from 1998 to 1999, and world-wide electricity from wind generation has increased by 39% in the same time frame. In contrast to traditional fossil fuel sources, wind generated electric power showed a 25% increase in 1997, compared to a 1.4% increase in oil consumption.[230] Likewise, in 1997, BP invested $1 billion and Shell $500 million in wind, solar and other renewable energy resources. These statistics further encourage use of the same rural land for farming and electric production. And most of the added income produced from the combined use of cropland stays in the local community, offering a sustainable means to revitalize rural communities.

Likewise, electric power from captured solar energy averaged 15% annually from 1990-96, and grew by a whopping 43% in 1997. The cost of solar cells has dropped 95% since the 1970s, and should drop another 75% during the next decade.[231] These prices would be commercially competitive with fossil fuels if there were no subsidies to coal-burning utilities. Additional incentives include the fact that fluorescent light bulbs use less than one quarter as much electricity as traditional incandescents.[232] Production has soared eight-fold in the past nine years. The one billion bulbs today lower electricity needs equal to the output of 100 coal-fired power plants.

And, what about steps directed at correcting present environmentally harmful activities. The world market for pollution controls, waste management, recycling, energy efficiency and the like (eco-technologies) should top $600 billion in profit by 2001, or more than the global aerospace or chemical industries combined.[233] Utilizing eco-technologies already available, we could enjoy twice as much material welfare while using only half as many raw materials and causing only half as much pollution and waste.

Believe it or not, the benefits far outweigh the costs. For example, during 1970-91, direct costs in the U.S. electric utility industry rose from implementing pollution prevention measures by 8%. But adding the indirect benefits of cleaner air showed that total utility costs actually declined by 15%.[234]

SUSTAINABILITY ANALYSIS

Do unto others as you would have them do unto you. Each individual, whether they are personally empowered, or not, has a vested interest in helping their neighbors be the best that they can be, because we are all in this together. I must do it myself, but I cannot do it alone.[235] We must let our own lights shine and give permission to others to do the same...loyalty and cooperation instead of distrust and confrontation.[236]

In the Seattle Special Olympics a few years ago a most enlightening thing happened. During the 100 yard dash, one of the contestants fell down and began to cry. Everyone in the race stopped, came back to the fallen runner and helped him get to his feet. They all crossed the finish line together. What matters in life is not winning but rather making sure everyone crosses the finish line. The competitive theme that spurred so much has now turned on us, and this one glimmer of light from those we deem as "less than" teaches us more than any science or technology book could ever attempt to teach.

As sustainability is achieved and all its potential is considered, an integration of planning becomes an obvious step. The first part of the plan is putting our doubts aside. Western Union was sorely wrong about the telephone, stating in 1876 "This telephone has too many shortcomings to be seriously considered as a means of communication." Is it possible many will also make inaccurate judgements about sustainable development?

Equally important is to be sure we make every effort to fully assess the impact of our actions and technologies, well into the future. For example, consider this story published in an airline magazine.[237]

*   *   *

"A dangerous new universe of communication threatens to ensnare us, one so seductive and insidious that it must be recognized for what it is: a web that will trap us. A web that will do nothing less than change the boundaries of human decency, decorum, and behavior...change the way we speak, think, and act. And what is the spider weaving this snare into our homes and offices, which mesmerizes and even enslaves its users in a wider and wider web, even world-wider? It is Mr. Alexander Graham Bell's lethal instrument, the telephone."

"The truth is that this cold, impersonal new method of communication has the potential to change us from a nation of warm, gregarious souls, welcoming our friends across thresholds, into a nation of lonely, housebound hermits. Front porches empty as desperate families sit inside on summer evenings waiting for the discordant clanging of Mr. Bell's hellish bells. Streets and lanes are deserted as former strollers find themselves tethered to the box on the wall. Polite dialogue is replaced by the grunts of the cave-dweller since face-to-face encounters will all but cease. Loathsome legions of salesmen, ordinarily turned away at the door, breach the sanctity of home and hearth whenever they wish to force their wares on those who have unsuspectingly responded to the bell. Language will be replaced by penny-saving grunts and snorts and followed by the withering of punctuation and spelling and the crumbling of grammar and syntax. Even travel, the great brightener of human experience, will cease, another victim of this web as we might simply talk into the telephone and ask those already there to describe the delights of Paris or Parma, Cairo or Cannes. Perhaps we should be grateful these telephones are held in one place by their wires, lest those who have been ensnared carry these devices everywhere."

*   *   *

Does this sound like another web that society has now become very acquainted with?

With the doubters and under-estimators placed safely in the back of the audience, planning begins. Planning involves looking at what is in place, what types of systems are needed, and how the people of that area can best adapt. The aim of this integration is to carry on our lives without harming the environment we live in, while always leaving ourselves open to new discovery, and to eventually contributing to a divine balance between honoring nature and obtaining our own needs for financial, emotional, and physical security. There is safety in a more socially balanced society. Only under these conditions will we look deeper and discuss decisions and ambitions...especially when considering the effect of our basic needs on other life.

These basic human needs might best be defined by the economist from Chile, Manfred Max-Neef.[238] He divides human needs into nine specific categories:

- permanence/subsistence;
- protection;
- affection;
- understanding;
- participation;
- leisure;
- creation;
- identity; and
- freedom.

These are the basic satisfiers of most human wants and sustainability analysis inquires how these satisfiers can be made more effective, while at the same time not doing any harm to society or the environment. As a means of acting, guided by this thinking on human satisfiers, the scientifically agreed-to principles upon which The Natural Step is based[239] can be used to generate four personal and basic conditions, guides, or principles for sustainability that can meet basic human needs. They are as follows, reprinted after the ideas of Gips.[240]

*   *   *

Condition Number 1: NATURAL EARTH CHEMICALS - substances from the Earth's crust must not systematically increase in nature. We should use renewable energy and nontoxic, reusable materials in order to avoid the spread of hazardous levels of mined metals and pollutants.

Why?—Mining and burning fossil fuels release a wide range of substances that do not go away, but rather, continue to build-up and spread in our ecosphere. Nature has adapted over millions of years to safely tolerate specific amounts of these materials. But, plant, animal, and human cells don't know how to handle over abundances of lead, mercury, radioactive materials, and other hazardous compounds from mining, often leading to learning disabilities, weakening of immune systems, and improper development and functioning. The burning of fossil fuels generates dangerous levels of invisible pollutants which contribute to smog, acid rain, and global climate change.

Action:—We can reduce our overall energy use. We can drive less, car pool, use public transportation, ride bikes or walk. We can conserve energy through energy-efficient lighting, proper insulation, passive solar, and reduced heating and cooling. We can support a shift to renewable energy such as solar and wind power instead of nuclear, coal, or petroleum. We also can reduce our use of mined metals and minerals through recycling and preferably reuse. We also can avoid chemical fertilizers.

Condition Number 2: MAN-MADE CHEMICALS - substances produced by society, must not systematically increase or persist in nature. We could use safe, biodegradable substances that do not cause the spread of toxins in the environment.

Why?—Since World War II, our society has produced more than 70,000 chemicals, such as DDT and PCBs. Many of these substances do not go away, but rather, spread and bio-accumulate in nature and in the fat cells of animals and humans. Cells don't know how to handle significant amounts of these chemicals, often leading to cancer, hormone disruption, improper development, birth defects, and long-term genetic change.

Action:—We can use non-toxic natural cleaning materials and personal care products. We can decrease our use of plastics and reuse the ones we have, such as plastic bags, plates, cups, and eating utensils. We can stop using CFCs and other ozone-depleting substances. We can use safe, natural pest control in our homes, lawns, and gardens. We can support farmers in becoming sustainable and eliminating hazardous pesticides by voting with our food dollars for certified organic food and clothing. We can support the elimination of factory farm feedlots and manure ponds that cause air and water pollution.

Condition Number 3: NATURE'S PRODUCTION and DIVERSITY - the physical basis for the productivity and diversity of nature must not be systematically destroyed. We should act to protect our soils, water, and air, or we won't be able to eat, drink, or breathe.

Why?—Forests, soils, wetlands, lakes, oceans and other naturally productive ecosystems provide food, fiber, habitat and oxygen, waste handling, temperature moderation, and a host of other essential goods and services. For millions of years these natural systems have been purifying the planet and creating a habitat suitable for human and other life. When we destroy or deplete these systems, we endanger both our livelihoods and the likelihood of human existence.

13 Intensive corporate farming in action. A worker spraying massive amounts of pesticides on rows of tomatoes in plasticulture on the Eastern Shore of Virginia (USA).

Action:—We can purchase sustainably harvested forest products rather than destroying rain forests. We can reduce or eliminate our consumption of products that are not sustainably harvested, such as fish. We can shop with reusable bags rather than using more paper bags. We can decrease our use of water and use composting toilets that return valuable nutrients to the earth. We can fight urban sprawl and encourage the cleaning up of brownfields and other contaminated sites. We can safeguard endangered species by protecting wildlife habitat.

Condition Number 4: EFFICIENT USE OF RESOURCES - there must be just and efficient use of resources with respect to meeting human needs. We can use less stuff and save money while meeting the needs of every human on this planet.

Why?—Our society is extremely inefficient in making and using things. About 93% of what is made is waste and only 7% is the product. Paul Hawken estimates that the manufacturing of a 4 pound lap top computer uses about 40,000 pounds of resources.[241] The American Society of Engineering estimates that we are only about 2-3% efficient in our use of energy and resources. Using 1,000 throwaway plastic teaspoons consumes over 10 times more energy and natural resources than making one stainless steel teaspoon and washing it 1,000 times. Furthermore, the planet cannot withstand our continued level of consumption and population growth. The U.S. makes up only 4% of the world's population but consumes about 25% of its resources and produces about half of its waste.[242] People living in the lowest 20% of the global population as classified by income, receive only 1.4% of the world's income.[243] Just to survive, they see no choice but to cut down rain forests, sell endangered species, over harvest fish, deplete coral reefs, and burn coal and other polluting energy sources. Buildings account for one-sixth of the world's fresh water withdrawals, one-quarter of its wood harvest, and two-fifths of it material and energy flow.[244] In addition, commercial buildings account for almost 17 percent of all greenhouse gas emissions.

Action:—Make business, government, and non-profits aware that to be sustainable we should achieve a ten-fold increase in efficiency. This will create a huge opportunity for saving money, creating jobs, and reducing waste as part of a new Industrial Revolution. We can make more effective use of our resources through reducing, re-using, recycling, and composting. We can redesign buildings, industrial processes, and products so that they can be safe and sustainable. We can encourage discussions among our fellow workers, friends, and families about what our basic needs are, ask if we really need more stuff, and design our workplaces, homes, and organizations in such a way to give us more of what we want (healthy, attractive, and nurturing environments and communities) and less of what we don't want (pollution, waste, stress, and expense).

*　　　　　*　　　　　*

The definition of a socially sustainable society is a society where you have stability, diverse dynamics, and a high degree of wisdom. The aim of sustainability analysis, following the "needs analysis" of human satisfiers from The Natural Step process above, is to encourage any project or activity to contribute to the development of a socially sustainable society. The strategic question for each person to ask themselves in thinking about making a decision or taking an action in the sense of The Natural Step, therefore is…are we violating these system conditions in our operation? Adding the four system conditions together equals self-sufficiency.

As we ready ourselves and our communities for a sustainable development movement we increase our ability to adapt and re-invent our self-sufficiency. We better our community by creating and maintaining a diversified social and economic base with locally shared ownership and access to basic human services. One example with grand vision for self-sufficiency is the mayor of a town who works to bring social balance and equal opportunity to his community by including Internet and cable access as part of the package a community member has with their water, sewage, and electric. At the same time he is supporting the farmers' markets, as well as a community bartering and trade system on a local radio station.

The more a community adapts to its particular circumstances, the more that community creates stability from inside out (which is what therapists have been saying for years). The degree of dependence by a community on outside needs equates to that community's degree of potential exploitation by outside forces. The more a community increases it ability to meet the majority of its own needs within itself, instead of being dependent on outside sources, the more stable that community is.

LOCAL MARKET ECONOMY

As a community looks at being more self sufficient it is essential to look at the weather, the terrain, and natural life of that community first. It is not a good idea to build an igloo in Arizona. The bio-region is a critical concept in evaluating our ecological impacts. As we look at the easiest ways to live within what our immediate environment can provide for us, we must adjust to the circumstances in which we live. How can we heat or cool our homes without distant polluting power plants? Or, how can we reduce our garbage output so we are able to safely dispose of it locally? How can we fill the shelves of our grocery stores with local products—perishable and non-perishable items?

Forcing the issue entails that we the individual must take back control of our own life. We can gain this control by means of supporting ourselves through local community. In encouraging our regional economies we must look for the hidden assets of the community on which we can build this new, smaller, self-sustaining economy. A stable economy builds on what is already there…a place where we grow our own vegetables and support the local farmers to supply us with our meats; and a place where we know who made the clothes we wear; and a place where we can contribute in a manner in which we see results within the space in which we live.

There are many questions we as individuals and community members could ask ourselves, each other, and the leaders of our communities. As communities practice sustainable development a collective vision of sustainability within a well-defined bio-region is essential. Otherwise, the communities are no more than economic colonies for the national and international economic markets.

Buying locally grown foods for example, is a fabulous way of promoting sustainable agriculture. Usually when we think of sustainable agriculture, we often think only of soil conservation, rotational grazing/growing techniques, and safe pest control measures. Sustainable agriculture also invites consumers to get more involved in food production farming by becoming active participants in the food systems. In North America a typical food item travels 1,000 miles before it reaches our meal plates, thanks to the artificially cheap cost of transportation with its absurdly low price of gasoline. If gas were to be realistically priced, people in Wisconsin might not be able to afford consuming quite so many strawberries in January.

A key goal is to understand agriculture from an ecological perspective and balance community and consumer needs. For example, the first reason to support locally grown and processed foods is economic. Money spent for food in the local area remains in the community and creates jobs for area residents. Another reason that regional food supports long-term sustainability is ecological. Local food systems allow urban and rural residents to build partnerships for healthy water, clean air, and safe food. When local farmers sell to consumers who live in the same community, they may be more sensitive to the environmental impact of production methods than absentee, corporate farm organizations.

To focus on just one example, consider the good old farmer's market. Farmers' markets resurrect the market town way of life, and make it easier for consumers to make the "right" choices. What they have done is grab an age-old concept that seems to strike a chord with the public...face-to-face contact between producer and consumer is the "best" model for sustainable development.[245] It is self-limiting, low-cost, with a small environmental footprint, high on social interaction, high on traceability, high on stimulating the local economy, etc. Other activities can also be built around farmers' markets (*e.g.*, cooking classes, information-giving, and awareness-raising for issues such as the concept of seasonality).

Local markets also work equally well in urban areas...for example, in run-down housing neighborhoods where there are few shops and residents can't afford transport to out of town supermarkets. They would welcome cheap, fresh produce. It also has the domino effect of local shops buying more from local growers.

Farmers' markets generally have three basic principles:

(1)   the producer of the product personally sells his or her own goods only;

(2)   goods are sold directly to the final consumer—that is, retail; and

(3)   the goods are produced from within a defined local area, say a radius of 40 miles.

Farmers' markets are thriving—new one's are setting up all the time. The public seems to be willing to spend increasing amounts of money at these markets. Why? It can only be because of those unique principles. There is a perception that the carrots are different, because they are marketed under the "home brand" and therefore have extra value attached to them, namely the social/environmental factors that are so desired.

Several regions of the U.S. and a number of European countries have been experimenting with the concept of Consumer Supported Agriculture (CSA) in recent years. This community practice is also becoming known as "subscription farming"[246] because of the way the process works. Consumers pay up front for products, as you would a subscription to a magazine. This supports local farmers by sharing in their economic risks from the beginning, so they are not surprised by the market situation when they go to sell their crops. The purpose of this marketing approach is to encourage local consumerism of commodities and cooperative buying strategies, as well as to offer more stable markets for small family farms while they provide lower-cost agricultural goods.

This direct farmer-consumer relationship encourages people in communities, neighborhoods, or even city block groups, to eat local, eat fresh, and eat seasonal. It also helps build a stronger sense of community and eliminates the environmental impacts of transporting food commodities long distances. CSA further allows one to know where their food really comes from, how it is grown, and what kinds of "growth" products (fertilizers, altered genetic makeup, etc.) go into producing it, all of which translates into a very important global issue now involving the whole idea of food system security.

Of the more than 1,000 farms across the U.S. that have experimented with this approach toward a more sustainable lifestyle, they have discovered a means to promote local consumerism that can lead a region toward self-sufficiency and enhanced agricultural economic stability. Decreased product costs for agricultural commodities and cooperative buying agreements also have been found to lessen the burdens to low-income populations in regions where this approach is being tried. The major barriers to this kind of lifestyle change relate to the willingness of small family farmers to make changes in their way of doing business and equally the ability of the rural public to develop new attitudes toward their purchase of local agricultural products.

In summary, local economic self-sufficiency is crucial if we are to reduce overall resource use because it cuts travel, transport, and packaging costs, and the desire to build freeways, ships, and airports, etc. It also enables communities to become independent of the global economy and community members to become more in tune with each other.

COMMUNITY, INSTEAD OF GLOBALIZATION

There is a call for our needs and our lives in general to evolve closer to our homes. The characteristics of our day to day activities are more secure if they are of a smaller-scale. The ability to know that our basic needs can be met locally brings much peace. It is very difficult for a person in New York City to fly to Greely, Colorado in order to meet the guy who he is buying his meat from. Especially because of the fact that it is not a person, but a corporation. A corporation who has shown over and over again a total disregard for life (human and animal). A corporation that simply no longer can be trusted to any degree.

But the guy from New York only sees the pretty package with the happy cow, or fluffy white chicken. He assumes his food comes from the kind of farm we all read about as children. But his food does not come from a farm. What he is buying comes from a place of inhumane degradation. Thus, the question of the day is: Would he still buy that product if he knew all the devastation it caused on the way to his table? If we were not once or twice removed from our products, a lot of production would be done very differently. The corporations count on our not knowing how they are serving up the mass quantities of meat, vegetables, dairy products, or whatever. And, it is this exact reason, or at least one of them, why we should localize our economies. It is this exact reason why the guy from New York, or the woman from Chicago, or the family from rural Maine must seek ways to meet their needs locally or regionally.

Though community-based technologies and economies at first glance seem too difficult, and seem to represent movement in a direction opposite to the popular globalism we hear so much about today, this movement really does not affect the global balance that is evolving. Global balance is about equal opportunity, health care, and education, as well as about an evolving consciousness for interconnections between humans and nature. Global balance is **not** about how far we can ship various supplies of anything and whether it still arrives intact.

Ecological consciousness has most strongly developed within a "minority tradition" that includes tribal cultures, utopian communities, and many religious traditions, such as within the Benedictine Order, Taoism, and Buddhism. And, it is truly on a small scale that individual responsibility begins. Participating in local community democracy is where we begin to take our power back. Technology can still exist. We can still chat on the Internet with the charming, deaf Russian and laugh at his spelling and perception of life as he has just moved to L.A. and as he discusses the humor he finds in our customs. We can still order books from Amazon.com and we can still learn more from the World Wide Web than sometimes we want to. And, we can still consider the value of technologies, especially if they are indigenous in development.

But when it comes to our day to day basic needs, we will find security in knowing the person that is growing, making, or building for us. And, security that we too have something very tangible to offer our local community, be it leadership, quiet agreement, or apples from our orchard.

An absentee landlord is an outdated concept that is void of responsibility and darkly liable in this day and age. What is needed is to take a look at the world that is immediately around us, adjust to what it offers, and develop our communities within this local framework. We live amongst life, it is definitely time we worked with it, verses against it or aside from it. People are committed to "re-inhabiting" and restoring that particular ecosystem and developing a renewed sense of place, in a way that will promote true self-sufficiency.

APPLYING A MEASURING STICK CAN HELP

When we consider any fact about our surrounding world, we must consider it in light of two basic issues: how do we affect it and how does it affect us. Indicators are our link to the larger world. They offer a measuring stick that can sometimes condense the world's enormous complexity to a manageable amount of meaningful information...to a small subset of observations informing our decisions and directing our actions. The rate at how things change in these chosen indicators provide clues about changes in a total system. The indicators we choose to tell us about the state of affairs for any system must be compatible with the pace (rate of change) of the system to provide us with timely information about its dynamics.

Two kinds of indicators can provide us with valuable information—the current state of our own life and the state of the life that is around us. For example, consider the relationship in the two groups of airplane instruments (indicators) that provide information about (1) the current state and functional integrity of the airplane itself, and (2) its position and heading with respect to the intended destination of the plane. If both groups of instruments are not working well and providing an indication of accurate operation, the plane will not reach its intended destination. In the case of a consumer of natural resources the "indicators" are how much destruction a behavior or action has caused to the world and people around us, and to ourselves. If we are hurting ourselves we will obviously not survive. Likewise, if we are hurting the environment around us we will probably not survive either.

In view of the dynamics of real systems, it is essential to focus on those things that provide early warning of impending threats, leaving enough time for adequate response. This requires having reasonably good understanding of the systems involved under different conditions. There must be enough time for an effective response. The time it takes to get an effective response under way must be less than the time actually available before complete system failure. If the rabbit outruns (response) the pursuing fox (threat), it will be safe; if not, it will be eaten.

In analyzing systems, we often attempt to condense large amounts of information to at least a recognizable pattern of a few understandable indicators. It is important that we use several different kinds of indicators, however, to obtain a "good read" on any particular situation, rather than relying solely on one indicator. Consider for example, a person's wealth indicating a long-lasting, happy lifestyle, where the wealth is measured only by their bank account balance. What happens when the bottom falls out of the stock market?

Actually we all carry out this kind of process of using indicators in our daily routines. A smile signals friendliness; a gray sky, possible rain; a red traffic light, danger of collision; the hands of the watch, the time of day; a high body temperature, illness; rising unemployment, social trouble. The more complex our own world, the more indicators we have to watch. We have to consider indicators that provide information about current and possible future developments.

Likewise, the identity of vital signs (indicators) of the health of a community's natural infrastructure (surrounding environment) and the understanding and careful monitoring of those aspects of the system can give people a clear picture of the systemic overall health of the ecosystem. If the thermometer indicates a temperature of 103 degrees F in our child, we know that his/her health is troubled. How do we monitor the health of the natural systems we rely on for survival?

Clear goals, meaningful policies and standards, identified programs, and agreed upon indicators of progress are necessary if advancement toward humanity's common future is to be charted and regular corrections to that course determined and carried out. In addition, the peoples of the world will be much more inclined to support policies and programs that emerge from the development of indicators based on personal standards they personally value than they would be to endorse objectives and initiatives which are based on a purely scientific or material conception of life. The use of these measures could, thus, help to transform not only the vision but the actual practice of development.

In our national accounting system for example, which we discussed extensively earlier, our Gross Domestic Product (GDP) counts all production of commodities and services and income of monies as "goods." In tallying up this GDP we never subtract any bads. In sum, the nation's brightest economists maintain our national accounting system with a calculator that has a plus key, but no minus key. Therefore, we have no way of knowing whether the costs of economic growth have exceeded the benefits. Not really a very good indicator! The nation's economists, politicians, and business leaders simply assume that if GDP is rising, our standard of living is rising too. The problems here are obvious when denial is removed.

In contrast, consider the fact that even though the warning signs are enormous, our willingness to accept the significant amount of chemicals in our environment is amazing. Like the alcoholic who experiences blackouts (periods where the memory stops imprinting but the person is awake and functioning), people daily witness dirty air and water, as well as animals with tumors and illness as a measure of toxic chemicals in our environment. The alcoholic tends to rationalize the blackout as normal, and society has ways to rationalize chemical use and destruction. Ironically our way of dealing with the economy and all its varied indicators, such as the GDP as a super indicator, show our incredible ability to adhere to certain kinds of warnings. When the Chairman of the Federal Reserve, Alan Greenspan, speaks for example, we all listen.

If indicators can work for economics, why can't they also work for our environment? For each of the presently identified 148 toxicants present in our environment, the U.S. Environmental Protection Agency (EPA) established a "benchmark" level that the agency considers safe. Eight (8) of the 148 industrial poisons have exceeded EPA's benchmark safety levels all of the time in all 60,803 census tracts monitored in the US in recent years.[247] All 8 are carcinogens, that is, they are known to cause cancer. Even more scary, according to careful measurements of human bones, pre-Columbian inhabitants of North America thousands of years ago had average blood lead levels of 0.016 ug/dL—625 times as low as the 10 ug/dL now established as "safe" for children. On the face of it, it seems unlikely that levels of a potent nerve poison 625 times as high as natural background—or even 300 times as high as natural background—can be "safe" for children.

Indicators and our lack of tolerance to change tend to show we have passed normal limits. They also tend to say we have exceeded safe amounts and have entered a zone of risk to our health. Crossing that line is very important, because shortly after crossing the line, where it takes more and more and more to feel normal, we cannot deny the physical consequences of our destructive patterns. And, the rabbit will definitely be eaten by the fox!

"Be conservative" is the message of the day. Our application of a measuring stick that includes indicators we are confident in can offer one means of being conservative. Let's avoid paralysis by analysis, but rather err on the safe side, advocating precaution where potential danger looms—even if we do not know the exact nature of the hazard. We may not know exactly how nature works, but by using fundamental laws and known relationships (indicators) we can calculate useful estimates of human demands. They may not be precise enough for managing nature, but they do provide challenging guidelines for managing ourselves in an ecologically and socially more responsible way.

The concept of erring on the side of caution, however, can often be difficult to grasp. A giant step was taken lately in the airline industry. Remember the crash of Alaska Air 261 into the Pacific Ocean off California in January 2000? Following this crash the Federal Aviation Administration ordered that all similar planes in U.S. airline fleets be immediately inspected for problems with their rear stabilizer systems. More than 2,000 planes were affected by this order over the duration of a week, probably causing a significant loss of revenue to the industry. But the death of 88 people aboard Alaska Air 261 warranted the caution that was being demanded to deter further disasters.

Ironically, more than 500,000 people in the U.S. die of cancer every year (as measured in 1992) and more than 30,000 of these deaths can be attributed to environmental causes.[248] And yet this is not enough deaths to warrant erring on the side of caution with regards to determining the effects of many chemicals we consistently emit into our environment with at least a suspicion of cancer causing effects. What makes an airplane tragedy with 88 deaths take precedence over the loss of thousands of people each year to environmental poisons we don't know enough about?

THE MONHEGAN ISLAND STORY

There is a wonderful island 12 miles off the coast of Port Clyde, Maine (USA) called Monhegan Island. In fact, this island has been called a microcosm of what sustainability should be about.[249] Monhegan's geography, its resource-based economy, its sense of place, and its long history have contributed to its sustainable way of life. In the sense of a microcosm that can be reproduced elsewhere, Bernard and Young suggest that Monhegan represents a path other places with tourism and resource-based economies can emulate.

So what's behind the success of self-sufficiency and ultimately some degree of sustainability that Monhegan Island has achieved through the years? First, and foremost, in the early 1950s Ted Edison, Son of inventor Thomas Edison, purchased and legally preserved 65% of the Island as "undevelopable" for ever because of his "love, respect, and admiration for the land." Since this initial publically accepted commitment, that now leaves the island more than 70% forested, the people of Monhegan, respecting private property, business, a long tradition of self-rule, and honoring individualism, developed important partnerships with state and federal agencies to protect their shoreline for fishers, save distinctive natural areas, avoid mass tourism, and prevent contamination of offshore waters as well as their sole-source freshwater aquifer.

The people of Monhegan live a good life in a difficult place. Irrespective of the hardships of living out in the north Atlantic, there is full employment, a modest tax rate, an excellent school, above average family incomes, and a general sense of well-being. This is not to say that there is not significant disagreement and sometimes conflict. These people have diverse opinions on how they manage to walk the line between individualism and community. But at the same time, in the end they always find themselves working towards the common good of the community.

Here are a few examples of how they have worked towards a self-sufficiency that in the long run is leading them down the path of sustainable development. They certainly know how to set limits, which is an exception in most New England coastal fishing towns quickly deteriorating because the fish stocks supporting people's livelihoods are disappearing. In the 1940s lobstermen on the island persuaded the Maine State government to allow a closed season, from July through December, on lobster fishing on the Monhegan fishing grounds, approximately 2 miles in all directions offshore from the Island. This was accomplished with a real sense of pride and many community members believe that their setting of limits explains the success and continued interest of Island lobstermen (and women) to this day. Lobstering is a very hard way to make a living and by these folks only doing it six months a year, they maintain a renewed sense of their culture and heritage every time the first of January comes around each year. These people do not celebrate New Year's Eve, they celebrate the first hour of the new lobster season.

Because of its wildlands, coastal setting with towering cliffs, and historic New England charm, Monhegan like so many other areas of coastal New England, through the years has also become a mecca for tourists. Again, these insightful people realized the meaning of limits. While other parts of coastal New England have been overwhelmed with tourism development, Monheganites have established mechanisms to limit this impact. In addition to nurturing a thriving tourism industry that supports the residents during the lobstering off-season, the Island because of its natural beauty has also attracted artists from around the world to spend up to six months doing their arts and crafts, selling their products to tourists, and providing significant additional revenue to the community's tax base. In turn, the presence of the artists for up to six months has enhanced the draw of tourists in a consistent and manageable flow.

This is not to say that there are not rough times in the Island community. The sole source of water to the town's people comes from a bog and fresh water marshland that the village homes (presently about 80 full time residents) ring the perimeter of. A number of years ago, with growing numbers of tourist visitors and increases in home luxuries requiring more water and the production of more wastewater, the town realized they were poisoning their only source of drinking water from their practices. Although it took some time to reach agreement among all the island's permanent residents, in their tradition of always "taking care of things themselves in a time-honored and civil way," they found the means and the will to amend recommended land-use ordinances and institute ways of diverting sewage so that their central meadow watershed would be protected and their only source of water maintained uncontaminated. In this process, residents admit that some were only looking after "their own skins," but not without enlightenment of the bigger picture and the destiny of their future.

As Ted Bernard so aptly states,[250] on Monhegan Island "is a lifestyle and loosely clustered set of natural resource attitudes and practices that in total have profound lessons to teach." The people's love of their island and their sense of place, and community, will hopefully be able to sustain their children and their children's children in a way that the future does not forget the past and recognizes that humans are just one piece of the puzzle.

COMMITMENT TO LIFESTYLE REDESIGN

While rivers, seas, and forests are generally healthier in some parts of the world, lifestyles in these areas are in many cases still grossly unsustainable. Democratic nations may have reduced local problems, but by importing fish, timber, food, and minerals from the rest of the world and exporting pollutants such as carbon dioxide, they are doing more than their part to spoil the global commons. Strategic reorientation of organizations toward sustainability, however, is beginning to happen. Businesses, governments, community groups—everyone is struggling with what it takes.

One community program that has been practiced in Portland, OR has neighbors in urban block areas working together for a more sustainable lifestyle. Establishing Eco-Team Programs based upon sustainable lifestyles[251] was driven by the motivation for concern about a community's environment and a commitment to making sure there would be enough resources available for their children to live decent lives. To this end people set about reducing their resource use and making more "eco-wise" purchases. They made these lifestyle changes in cooperation with their neighbors, whom they had gotten to know and liked in the process.

These people were literally able to redesign their lifestyles for resource efficiency. For example, if somebody had to go to the store they contacted their neighbors to see if anybody else needed anything. This minimized the use of transportation within the neighborhood. Likewise, the recycling that one household did, all in the neighborhood did, which made it not only more effective but also easier on the individual household with the help of others. The statistics tell the story: 50 percent less garbage sent to the landfill; 34 percent less water used; auto emissions reduced by 20 percent; 9 percent less energy used; and an average household savings of $265 per year.[252] These community lifestyle campaigns are now underway in major metropolitan areas as diverse as Chattanooga, TN; Kansas City, MO; Columbus, OH; Issaquah, WA;, Philadelphia, PA; Bend, OR; Madison, WI; and Minneapolis, MN.

*   *   *

The City of Malmo in southern Sweden is building a new residential and commercial district that is intended to be a leading example of sustainable development in a high-density urban environment. The district will be entirely powered by locally produced renewable energy, using energy from solar, wind, and wastewater sources, as well as a range of energy efficiency measures.[253] Also, householders will be given the opportunity to track their own energy consumption.

Although the district will be connected to the regional electricity grid, it aims to have zero net energy consumption because of reliance on all the peripheral, natural sources incorporated into the village design. A heat pump will be used for heat generation, with seasonal storage in the limestone aquifer under the site's development setting. Heat will be stored during the warm months and pumped up and included in the energy system during the winter. By extracting heat in the winter from the warm underground aquifer, there is also a natural cooling of the aquifer which can then be used to cool buildings in the summer.

In Malmo, the rental car firm, Hertz, has agreed to organize a car pool, and give 30% cheaper rates for the use of electric vehicles. Some parts of the urban development, however, will be closed to traffic, with a network of pedestrian and cycle paths, and environmentally compatible public transport.

Wastewater will be used as a raw material. Phosphorus and nitrogen will be recovered and used in agriculture, and methane will be used for power generation. Rubbish will be sorted for recycling, with some fractions also used for methane production. All buildings will be constructed using environmentally compatible and resource efficient products and materials.

*   *   *

As in Malmo, it is not too late to change courses, take responsibility, and build a society that is both environmentally sustainable and industrial everywhere. Kalundborg, Denmark, is a blueprint for "industrial ecosystems" with 16 materials and energy exchange projects.[254] For example, an oil refinery (a) employs waste heat from (b) a power plant, and provides sulfur to (c) a wallboard producer. At the same time, the power plant (b) supplies steam to heat water for (d) fish farming, and warms (e) greenhouses and homes. The complex has greatly reduced its use of oil, coal, and water, as well as its emissions of carbon dioxide and sulfur dioxide.[255]

It is not too late to save those kingdoms that are teetering on the edge of extinction. It is not too late to show respect for these kingdoms, their castles, and all the life therein—seen and unseen. It is not too late to build a world where the air is safe to breath, water is safe to drink, and resources are shared among all the world's people. And, it is not too late to build a world that we recognize as the one we hope our children will inherit. We can work side by side, hand in hand. Achieving a sustainable lifestyle opens people to a greater compassion for one another as it helps people understand the consequences of their choices within the context of nature's impartial law of cause and effect.

Chapter 12

Sustainability in Motion: Some Solutions

"Just as the hand, held before the eyes, hides the tallest mountain, so does our ordinary way of seeing hide the many wonders of which the world is full."

Dr. Jack Vallentyne

Modern society faces the undoing that once brought down seemingly invincible civilizations in the past. Then, the collapse was comparatively local in scale. Today it is global. The decisions all of our global society makes over the next two decades are likely to dictate whether or not the Earth life-support systems are sustained or become irreversibly impoverished. Without careful thought about deeper values and goals, as well as appropriate policies and strategy, the best endeavors are likely to go round in ever decreasing circles. There is also danger in seeing individual issues in isolation rather than as aspects of one general systemic crisis, with related causes and linked solutions.

In these complex times it is hard to sustain individual involvement without the deep commitment that fuller awareness and understanding can bring. For example, as we become more aware that our consumption practices are more than the environment can tolerate, and as we understand that we can conduct our lives differently, we ask: "What is the responsible action?"

INDIVIDUAL ACTIONS: AWARENESS...UNDERSTANDING...MOTION

This section is dedicated to the various life sustaining elements of our Earth (air, water, land, food), as well as those components of society (chemicals and corporations) most responsible for affecting these elements. The following clearly identifies what each is, how it is affected by present societal actions (or non-action), and discusses potential solutions to consider as means to correct insults occurring to Earth's various life-sustaining gifts.

Basically we are building on the process of *Awareness*, *Understanding*, and *Motion*, a means of looking at sustainability issues we established at the start of this book, to be consistent in addressing the status of the modern day environment that both is affected by society and in turn affects society. Awareness, understanding, and motion can lead to a sustainable future. Once people have become aware and knowledgeable, they may be ready for actions, but they need and want to know what they can do that will make a difference. We hope that you will find our numerous suggestions for motion helpful in developing your strategies to act.

THE AIR

Awareness: Air, the oxygen in it, is a simple, well known, basic life support component. Without oxygen we suffocate quickly. In polluted air we suffocate slowly—no ifs, ands, or buts about it. No way to debate this little factor away. We know that pollution in the air kills. We know when we are breathing polluted air. We see the smoke come from the factories but support these factories as we buy their products. There are other ways to produce without that smokestack going full blast day in and day out. That factory can be energized by other means.

- Children in the largest cities around the world experience air pollution levels two to eight times above the maximum World Health Organization (WHO) guidelines and more than 80% of all deaths in developing countries attributable to air pollution-induced lung infections are among children under five.[256]
- More than 220 million Americans breathe air that is 100 times more toxic than the goal set by Congress 10 years ago.[257]
- For 11 million people the cancer risk from their neighborhood air is 1,000 times greater than Congress's goal.[258]
- Americans generate 140 pounds every day in climate-damaging greenhouse gases.[259]
- Global emissions of carbon dioxide reached a new high of nearly 23,000 million tons in 1996…nearly four times the 1950 total.[260]
- The average American is responsible for about 20 tons of carbon dioxide emissions per year, a far greater volume per capita than that of any other industrialized country.[261]
- If every household in California replaced 4-100 watt incandescent light bulbs with 4-27 watt compact fluorescent light bulbs, burning on average 5 hours per day, we would save enough energy to shut down 17 fossil fuel energy plants.[262]
- The annual electric bill to operate all the exit signs in the U.S. is $1 billion.[263] Imagine the carbon dioxide **not** emitted from fossil fuel power plants if these signs were solar powered.

*14 Factories are an extremely large source of air pollution and gases that contribute
to the greenhouse effect around the world. In addition, automobiles made by these
factories are an equally important source of air pollution.*

- If every household in California replaced 1 average-flow showerhead with an energy saving showerhead, we would save 1.3
  kwh per day per household or enough energy to eventually
  shut down 15 fossil fuel energy plants.[264]
- Most of the gases responsible for stratospheric ozone depletion
  are produced by human activities and are not naturally occurring in the atmosphere.[265]

Understanding: Factories are a large source of air pollution, but there are many polluters. The power companies, those cars blowing black smoke, sports utility vehicles, fires of any kind, and the burning of forests to create grazing land for cattle in South America (mainly for American consumption). When you make a difference others will follow. Look at the energy you use in your home, your work, your school, your church, and anyway else you tend to spend time. Ask, "is this efficient?"

- The nation's top 100 power companies are responsible for 90% of the pollution-causing emissions of nitrogen, sulfur dioxide, and carbon dioxide.[266]
- Every day on average, American farmers will pump an estimated 15 million tons of carbon dioxide into the atmosphere.[267]
- The average car now going to the junkyard gets better gas mileage than the car coming off the assembly line. But, only 5% of the people in the U. S. commute by bus or rail, down more than a third since 1970.[268]
- Every time you drive 20 miles you emit toxins into the air. Consider this amount times the 150 million Americans who also drive this distance each day.
- Polychlorinated compounds like dioxin can be formed by burning common household trash at low temperatures.[269] A family of four burning trash in a barrel in their backyard...still a common practice in many rural areas...can put as many dioxins and furans into the air as a well controlled municipal waste incinerator serving tens of thousands of households. The trash burned in these measures included newspapers, books, magazines, junk mail, cardboard, milk cartons, food waste, various types of plastics, and assorted cans, bottles and jars.
- An average of 14 of the total outdoor concentrations of 148 toxic air contaminants for all population census tracts in the U.S. exceeded their estimated safe "benchmark" concentrations for humans. Approximately 10% of all census tracts had

estimated concentrations of one or more air pollutant carcinogens greater than a 1-in-10,000 risk level.[270]
- In a single day the world's people will release about 1,800 tons of ozone-depleting CFCs to the atmosphere.[271]

Motion: The little things add up so it is the little things that will make a difference. Once you start with a small behavior change you have much less tolerance for the more obvious polluters. Thus, you may well be the one that gets the ball rolling to change your local factory's bad habits. As you realize that conservation is the key, you will drive less, car pool more, and make your trips well worth your time. Find our what kind of waste is emitted to heat or cool your home. Change the way you heat your home if these facts alarm you. Do it differently. Boycott those smokestacks.
- Insulate your home; build environmentally smart.
- Use florescent light bulbs and conservative appliances, thus reducing waste.
- Get those solar panels, thus reducing energy consumption and also reducing waste generation from fossil-fuel power plants.
- Get that solar water heater, thus reducing energy consumption and therefore reducing waste emissions from your personal gas burning.
- Get that solar refrigerator, thus reducing energy consumption and therefore waste.
- Conserve, conserve, and conserve.
- Wind energy is phenomenal—check into it. Imagine a windmill for pumping water.
- Bike more drive less. Maybe even walk once in a while.
- Help restrict the overall population of motor vehicles.
- Stop consuming products that through their production emit toxic chemicals into the air. If you don't know what these products are, find out.

THE WATER

Awareness: We are literally running out of water. There are fewer and fewer fresh water sources. In addition, most water supplies have some level of chemical poisoning. Wastewater treatment plants pump more and more nitrogen and phosphorus into water ecosystems, fueling the growth of unwanted aquatic plants and depleting oxygen which kill the fish. We waste so much water and we poison so much more.

- In 1996, 263 million tons of nitrogen and 18 million pounds of phosphorus ran into the Chesapeake Bay.[272]
- Fish advisories have become a standard practice of the 1980s and 1990s. Due to our polluted streams, rivers, lakes, and oceans, our fish are now so heavily laced with PCBs and DDT we can no longer eat them.[273]
- The average U.S. household uses 146,000 gallons of water a year, of which 42% is used indoors and 58% outdoors.[274]
- Currently a large proportion of the world's population is experiencing water stress[275] and rising population demands for water from irrigation (70% or all water uses), industrial (20%), and residential (10%) uses greatly outweigh greenhouse warming affects on world water supplies.
- The number of people living in water stressed countries is projected to climb from 470 million to 3 billion by 2025 and many countries do not have a coherent national water policy.[276]
- In arid regions of the world, which supply 30% of the world's food, irrigation has declined during the past decade from water shortages.[277]
- Satellite images show springs, lakes, and rivers, drying up throughout the northern half of China, while a government assessment reports that the water table under much of the North China Plain, a region responsible for nearly 40% of China's grain production, has fallen an average of 1.5 meters each of the last five years.[278]

- Mexico City is sinking as residents pump up the water beneath them. Elevated train tracks, built flat in the 1960s, look like roller coasters now.
- The Ogalalla aquifer that waters one-fifth of all U.S. irrigated land is overdrawn by 12 billion cubic meters per year, a problem that has already caused more than two million acres of farmland to be taken out of irrigation.[279]

Understanding: Should water sources be depleted, there will be nowhere to turn. Social chaos will rein, money will not buy a way out. Should our oceans and coastal waters continue to deteriorate more than 60% of the world's population that lives on the coasts of our oceans will be in jeopardy. Protection and conservation are needed now.

- If present consumption patterns continue, two out of every three persons on Earth will live in water-stressed conditions by the year 2025.[280]
- The average North American consumes over 170 gallons of water per day, more than seven times the per capita average in the rest of the world and nearly triple Europe's level.[281]
- In contrast, the World Health Organization says good health and cleanliness require a total daily supply of about 8 gallons of water per person.
- Approximately two-thirds of residential interior water use is for toilet flushing and bathing.[282]
- In the last half century, land irrigation by using powerful diesel and electric pumps have made it possible to extract underground water much faster than it is replaced by rain and snowfall. And yet traditional spray irrigation loses about a third of the water to evaporation and winds before it ever reaches the plant roots.[283]
- Up to 90% of the water used to sprinkle lawns on a hot sunny day can be lost to the atmosphere through evaporation before ever reaching the grass roots.[284]

- Global freshwater supplies are being used up so fast that almost half of a billion people already depend on non-renewable sources. Water riots such as those in China's Shandong province in July 2000 will become more common as people struggle for control of dwindling water supplies.[285]
- According to a report out of Washington D.C. on July 20, 1999…increasing water shortages may lead to global hunger, civil unrest, and even war.[286]
- While the Syrians press for an Israeli withdrawal from the Golan Heights, water, not land is the crucial issue between the two countries.[287] The Golan Heights provides more than 12% of Israel's water requirements.
- In 1986 a study statistically linked children with leukemia in Woburn, Massachusetts to contaminated drinking water affected by a nearby waste site.[288]
- Brain, nervous system cancers, and acute lymphocytic leukemia represent the majority of cancers attaching children. Clusters of these cases are occurring in regions where drinking water has been contaminated by carcinogenic volatile organic compounds discharge by industry and municipalities into underground sources of drinking water.[289]
- A study released in 1995 has shown that in 29 cities and towns in the U.S. corn-belt herbicides in drinking water exceed federal safely levels.[290]
- Between 1976 and 1996, annual occurrences of harmful algae blooms—a leading indicator of health risks for marine animals and people—increased from 74 to 329.[291]
- Strandings of whales, dolphins, and porpoises thought to be linked to poor oceanic environmental conditions jumped from nearly zero in 1972 to almost 1,300 in 1994.[292]
- Mass fish kills and disease outbreaks went from nearly unheard of before 1973 to almost 140 events in 1996.[293]
- But the most frightening jump was in the numbers of human health problems traced to coastal waterways. These include

everything from swimmer's itch, to *Pfiesteria*-induced memory loss, to cholera. There were only two reported incidents in 1966, but 118 in 1996.[294]

Motion: There are many simple things we can do daily to conserve and not poison our water system. A very simple fact...we cannot live without sufficient, clean water supplies.

- Use water efficient appliances: toilets, dishwashers, clothes washers, etc.
- Think about conservation when you continually spray water over lawns or plants, and constantly wash vehicles.
- Don't run the water as you brush your teeth, shave, or soap your hands. All these little things add up. Think about these activities exercised by all people on Earth and do your own math.
- Don't use the toilet as a wastebasket.
- Repair leaky faucets.
- Position your roof and gutter downspouts so rain water runs onto the lawn or into the garden, not down the driveway.
- Check your water meter or bill to see how much water you are using. Each of us should be able to get by comfortably on 50 gallons per person per day.
- Don't use pesticides, herbicides, or fertilizers on your lawn, gardens, and plants because these toxins run right into the water system.
- When you see farmers spraying chemicals, make them aware of what they are doing. Begin a local movement to end the spraying of chemicals where you live, or you can bet you will be drinking them in your water supply.
- Think about the food you consume. Factory farms are not only horrendously inhumane places, they are also a main user of water and a main polluter of the water system. Fast food burgers have a very high price.

- Conserve, think, and challenge yourself and others as you begin to understand what is happening to our water.

THE LAND AND ITS RESOURCES

Awareness: As with all the basic life support components of Earth we are discussing here, land is essential for us to live, for us to grow food, for us to travel on, and where many other animals and plants call home. It also offers untold ecosystem services that directly benefit human life. Be aware that if the land has poisons used on it, it will indeed become toxic and dangerous—a living mine field without the mines, but still just as deadly (remember Love Canal?). Also, there is only so much of it. There are only so many forests, so many ecosystems that can support life and crops, and only so much topsoil. The land is precious. One of our greatest strengths as a country (U.S.) is the fact that we are the breadbasket for the world. But this has been abused with over consumption, poor land practices, and wastefulness.

- Americans consume 120 pounds, nearly their average body weight, every day in natural resources.[295]
- In an average day 72 million tons of topsoil are eroded.[296]
- According to the most optimistic estimates by the oil industry, oil reserves will last approximately 45 more years at current rate of consumption.[297]
- As U.S. cities grow their urban/suburban areas increased resource consumption and waste assimilation requirements of a defined human population or economy are demanding a corresponding increased need of productive land area (more land per capita is being used by populations settling in these urban sprawl areas). The conversion of land to urban uses has occurred at 2.65 times the rate of population growth since 1950 in the U.S.[298]
- The amount of farmland and forests turned into development each year in the U.S. is more than 1.4 million acres.[299]
- The total area of public roadways in the U.S. is about 15.7 billion acres, and parking and driveways total 7 billion acres.[300]

- Recent studies estimate that 140 to 180 square miles of tropical rain forest are cut down every day to provide wood and to make way for ill-fated farms and ranches.[301]
- Sierra Club research shows that 120,000 acres of wetlands are being lost in the U.S. each year, with about 99% of plans to destroy U.S. wetlands meeting with success.[302]
- The availability of land for other biological resources, besides humans, is declining as human population grows, causing a potential loss of 11% of all bird species and 25% of mammals.[303]
- In Illinois alone there are 1,500 hazardous waste sites that are in need of remediation—a list that does not include thousands of pits, ponds, and lagoons containing liquid industrial waste.[304]
- Westinghouse maintained a very lethal toxic land dump for years, as it negotiated with the EPA on clean up details. It recommended its staff wear respiratory protection while assuring the people of Bloomington Indiana there was no immediate danger.[305]

Understanding: Without healthy lands, countries would fall from power and experience all the chaos that goes with it. And, people would die. Look at Haiti. It use to be a rich, healthy environment until it was wiped out by deforestation: the same is true of Nepal and Lebanon. The land is what feeds us. Be gentle with it, and give back to it.

- If we captured just 1% of the sunlight that falls on the Earth's land daily as solar energy we would have more electricity than is generated by all other sources combined.[306]
- Despite growing concern about the environment, the average U.S. house size is 2,500 sq. ft., up from 1,900 sq. ft. in 1977.
- A bushel of apples that sold for $14 in the mid-1990s now sells for $9 and costs about $11 to produce in New York State. As a result growers have been leaving the business in droves and

developers are willing to pay handsomely for the apple orchards in the vicinity of transportation corridors to major cities, increasing sprawl rates on natural lands.[307]
- The amount of developed land in the U.S., excluding Alaska, totaled 98.2 million acres in 1997, or 5.1% of the continental U.S. land mass, an increase of 34% since 1992.[308]
- The lower 48 U.S. states have lost more than 53% of their wetlands since the 1700s.[309]
- Increasing incidence of serious flooding in the U.S. is the result of wetland development. Wetlands act as natural sponges and where you don't have wetlands you have quicker and more deadly floods from storms.[310]
- In a typical day on Earth, 70 square miles of land will become desert.[311]
- Some 20% of the world's susceptible drylands are affected by human-induced soil degradation, putting the livelihoods of more than 100 million people at risk.[312]
- Largely because of the rampant destruction of tropical rain forests, 40 to 100 species become extinct on a typical day on Earth.[313]
- Six billion animals are slaughtered each year for food. That equals 1.37 million pounds of manure that is created each year by these factory farms and directly placed in very toxic, open air lagoons.[314]
- In 1988 Union Carbide alone generated 301 million pounds of hazardous waste, an increase of 70 million pounds from 1987, a whopping 30.3% growth in waste generation.[315]
- There are an estimated 425,000 brownfields in the U.S., including 16,500 polluted sites in 126 cities covering about 47,000 acres.[316]

Motion: Once again, do not put poisons on the land. Do not cover everything with concrete and tar. Leave the trees. Their root system is vital to the water cycle. Leave natural systems in place. Tidying whatever area, as a whole, is not healthy for the land. Recycle, and find ways to use less, thus creating less garbage. Reuse all those plastic containers. Get the local grocer to have detergents, shampoos, and cleaning liquids in large containers so you can refill and reuse the original container. There are many products that this process will work for. There are many ways to pay respect to and protect the land.

- A tax on toxic dumping would discourage this anti-social past time.
- A tax on toxic raw materials would induce users to seek safer alternatives.
- Again look at what you consume, from T-shirts to meat. How much sacrifice occurred from the land for you to have all that you have?
- Plant trees and if you live in an area where there is a cutting down of forests, check into it. Find out more. Petitions can move mountains in stopping the cutting of forests that are vital to the health of the land.
- Do not litter your local landscape, or anybody elses.
- Recycle and watch how many bags of garbage you put out weekly—try and cut it in half.
- Think about the actual footprint that your house takes up of the surrounding landscape. Is it more than you really need?
- If you are moving in the future, think about living in the city instead of in a new suburb.
- Replace roads with homes, parks, and gardens.

THE FOOD

Awareness: The food we eat in today's corporately controlled farming and distribution industry is less than safe. From the genetically altered seeds, corn, and soybean products, all the way to the meat industry, our food has more chemicals, antibiotics, and bacteria that are not safe than people are aware of. It is very simple: the way we breed animals for food is a threat to the planet. And to top it off, clean up associated with the water and air pollution caused by factory farms are paid for by the taxpayers in the form of new water treatment facilities and doctor visits.

- Approximately 480 million of the world's 6 billion people are being fed with food produced with the unsustainable use of water.[317]

- A typical steer consumes 2 tons of grain to gain 400 pounds while at a feedlot.[318] How many people could this same amount of grain feed in under-developed countries?

- About 80% of cattle in he U.S. are routinely fed slaughterhouse wastes —the rendered remains of dead sheep, and dead cattle. The USDA banned this hoping to prevent Mad Cow Disease. So now millions of dead cats and dogs are purchased from animal shelters and fed to cattle, even though cattle are herbivores, eating grain and grasses.[319]

- The USDA followed obediently the lead of the FDA in its December 1997 decision to declare red meat irradiation safe (hitting meat with radiation to kill bacteria, killing many of the nutrients at the same time), despite a near complete lack of toxicological evidence supporting this position of supposed "safety".[320]

- Currently 50 million pounds of antibiotics are used in food production each year, with nearly half of them used to stimulate growth in farm animals. The rest are used to fight diseases due to the nasty, filthy, and inhumane conditions that these factory farms overtly maintain as if they were above all laws of God and man.[321]

- The use of bio-genetically enhanced corn grew from 80 million acres in 1997 to 400 million acres in 1998. This corn kills butterflies. That in itself seems to merit more caution in its use. And, it is shown that the corn and soybeans that are altered genetically to withstand more doses of the pest killer Roundup for gaining a higher yield, actually do not produce a higher yield.[322]

- The USDA tests one out of every 250,000 animals for toxic chemical residue, and even then it tests for less than 10% of the chemicals known to be present in the meat industry.[323]
- Some of foods most craved by pregnant women, such as meat and dairy products, may give their babies an extra dose of toxic pollutants by these woman increasing their intake of persistent organic pollutants.[324]

Understanding: The effect of using everything from hormones to irradiation is largely unknown. It is widely discussed yet literally unknown to many that we indeed are subjects of a huge science experiment. There is enough evidence to show the use of hormones in cattle do cause cancer. Likewise, the over-use of antibiotics is creating an intolerance of their effects. Thus, we could easily become ill by eating meat with a bacteria that is antibiotic resistant. There is more food than needed for everyone in the world, yet our consumption of meat, our wastefulness, and lack of information about what we eat continues to be a rampant threat to our health. Consider the degree to which factory farms pollute our environment while consuming huge amounts of water, grain, petroleum, pesticides and drugs. Add to that the fact that disposal methods on these factory farms are primitive at best and lead to disease and amazing environmental devastation.

- Hunger is really a social disease caused by an unjust, inefficient, and wasteful control of food.[325]
- Grain production, which supplies 80-90% of the world food, has been declining since 1983 and should alert us to the potential for future food supply problems and increasing malnutrition.[326]
- In the past 24 hours 45,000 people starved to death, 38,000 of them children. This occurs when the livestock population of the U.S. is fed enough grain and soybeans each year to feed the entire human population 5 times over.[326]
- Food irradiation has been banned in all countries except the U.S. It may kill many bacteria that get on meat at the massive

slaughterhouses and their filthy inhumane environments. In the process, however, it breeds a whole new strain of bacteria, that is very deadly.[327]

- Every week approximately 5 people die from eating a hamburger.[328]
- In 1988 human staph infections resistant to antibiotics were over 90%.[329]
- *E. coli* is a very deadly bacteria whose origins and existence is often correlated to dirty meat. Each year *E. coli* is responsible for 500 deaths and 20,000 illnesses. And, the number continues to grow.[330]
- The U.S. suffers from 2 million salmonella cases each year, killing 2,000 people. Again this goes back to the factory farm approach to raising animals and the filthy, inhumane conditions that exist in these animal warehouses.[331]
- Women who eat fish from Lake Ontario (North America) have higher levels of PCBs and pesticides in their breast milk than women who do not eat the fish.[333]

Motion: There are so many ways to obtain food beside the local super store with all it's pull and flash. Utilize your local farmer's market, start that rooftop garden. If you cannot have a garden, perhaps a community garden is possible. Where there is a will there is a way. America is the most over fed, yet poorly fed country in the world. What we eat is literally killing us. The choice is yours. Is meat so important as to gamble with our life so often?

- Grow your own vegetables.
- Learn to can and preserve vegetables and fruits.
- If you want to eat meat, find a local farmer that will raise grass fed, humanely treated animals for you.
- Ask about the milk: if it contains rBGH hormones, refuse to buy it.

- Use soy and rice based milk to tide you over until the corporations get the message that we will not buy tainted, cancer causing products.
- Know where your food comes from and how it was grown.
- Consume only products produced within your bio-region.
- If the American population alone would cut its beef consumption by only 10% per year, enough grain would be saved to feed 60 million people.[334]

CHEMICALS

Awareness: Be aware of the various chemicals you use. Be aware that if you are trying to kill something with whatever toxic spray, it probably will affect you too. All the chemicals we use in our houses, on our lawns, and in our gardens are made of toxins that end up in the water system. That is why every walking, living, breathing human being today has DDT flowing through their veins. And, in addition, be aware of what chemicals are in things that you decide to throw-away or in other ways destroy.

- One out of every three people alive today will experience cancer.[335]
- 1,400 Americans die of cancer every day.[336]
- Babies have been born without brains in Brownsville, Texas (remember there is here).[337]
- Some frogs have been observed in nature with only one eye or five legs.
- A recent five-year study has shown that children exposed to toxic pesticides have a noted decrease in mental ability and heightened aggressivity.[338]
- Fertilizer is not regulated by any federal guidelines. In Gore, Oklahoma they literally renamed their low-level radioactive nuclear waste "fertilizer" and sprayed it on grazing fields.[339]
- In a springtime ritual as old as the suburbs, millions of gardeners are spraying 2,4-D and other herbicides to cultivate perfect lawns, while keeping them free of dandelions and other weeds.

Rarely do these gardeners realize that they are usually applying more herbicide per parcel of lawn than farmers treating their fields. Dandelions have an Achilles' heel, however—a high demand for the mineral potassium. As a result, a lawn planted with grasses that don't require much potassium, such as bent grass, foxtail, and fescue, can be kept lush and green while dandelions remain in check, as long as potassium fertilizer isn't added. Many common lawn fertilizers contain a healthy dose of potassium, encouraging the growth of dandelions and, subsequently, the use of 2,4-D and other chemical herbicides to kill them. For many lawns, a fertilizer containing only ammonium sulfate or ammonium phosphate would be better.

- Exposure to hazardous chemicals has been implicated in numerous adverse effects on humans from birth defects to cancer. Global pesticide use results in 3.5-5 million acute poisonings a year.[340]
- Forty percent of U.S. waterways are considered too polluted from the dumping of toxic chemicals for safe fishing and swimming.[341]

Understanding: Understand that many chemicals regularly used in our society today actually alter cells, alter how individual cells send messages to one another, and in turn create new abnormal cells. The nervous system, the immune system, and the endocrine (hormone) system are all closely related and constantly communicate with each other. If any one system is damaged, it may adversely affect all systems.

- Multiple Chemical Sensitivity (MCS) is a term used to refer to a complex physical disorder involving adverse reactions to chemicals even in minute amounts that could ordinarily be tolerated–at least in the short term by individuals without MCS.[342]

- Toxic pollutants, specifically the toxic metals lead and manganese, cause learning disabilities and increase aggressive behavior, and most importantly, loss of control over impulsive behavior. Could this have anything to do with the increase in violence we see in our schools, or on our streets?[343]
- Pesticide use is up 3,000% since World War II, yet crop damage from "pests" has also increased 20%.[344]
- Scientists studying the 7,000 square mile "dead zone" in the Gulf of Mexico have placed much of the blame on fertilizer runoff from American farms in the Mississippi River watershed.[345]
- Pesticides, agricultural runoff, and animal byproducts are affecting fish, frogs, and the overall state of the Chesapeake Bay.[346]
- Ten percent of the world's pesticides are used on producing cotton. It takes a third of a pound of chemicals to make one cotton T-shirt.[347]
- Corporations have taken the stance that chemicals are safe until proven harmful. Thus, the burden of proof is on the potential victims. Again, just as with dairy products laced with hormones, corporations are using the general population for a huge science experiment.
- Pharmaceutical companies are the exception; before marketing to the public they must show their product not only safe but also effective, yet they still kill 140,000 each year.[348]
- An increasing amount of new development across the U.S. is occurring on former agricultural lands, and developers are being forced to reckon with the possible contamination of their sites from the use of now-banned pesticides.[349]

Motion: Although chemical production is controlled by large corporations, there are individual actions that will add up to balance the playing field. It is our responsibility as consumers to make sure that what we purchase is safe. No longer does the vendor take pride in their product.

- As for the corporations who overtly use toxic chemicals and illegally store and dump chemicals, boycott them.
- As for the stores who sell tainted produce and support the use of toxic pesticides and fertilizers, boycott them.
- As for the farmers who continually spray their fields with toxic insecticides, pesticides and fertilizers, boycott them.
- As for your neighbor who uses toxic insecticides, pesticides and fertilizers, educate them.
- As for your father, sister, brother, or whoever that uses toxic insecticides, pesticides, and fertilizers…educate them. Or if it is the holidays, and you're broke, boycott them too.

THE CORPORATIONS

Awareness: Corporations are out of control. They show little regard for individual life as well as provide false information, misleading the public. Granted, not all corporations are dishonest and destructive, but many are. No longer can we trust what is on label as accurately representing the full information about a corporation's product. And most importantly, no longer can we trust many products to not cause us harm. Corporations are more powerful than government at this time.

- The CEOs of the largest corporations earned an average of $7.8 million…326 times that of the average factory laborer.[350] Not a luxury easy to part with.
- In truth, the law of real people can no longer direct corporate actions. That is to say, corporations govern. Like Dr. Frankenstein's monster, the creation is now the monster of the creator.[351]
- The lowest 20% of working-age couples pay 12.4% of their income in state and local taxes while the top 1% only pays 5.8% taxes.[352]

- "Everyone should know the war on cancer is largely a fraud," according to Dr. L. Pauling, two-time Nobel Prize winner.

Understanding: There is no way that government can regulate quality products while being held hostage. It is up to "we the people" to demand quality, harmless products. The rBGH hormones in milk are shown to cause cancer, yet industry moved to ban labeling on milk identifying this hormone, and moved to outlaw milk producers from labeling their product as free of rBGH. There is little known about genetically altered foods and irradiated foods, yet industry plows ahead wanting to mass market these products. There are many examples of corporate tyranny. Again, ask questions, dig deep to find out what you consume and what the corporations are up to.

- From recent surveys and polls, as a result of unlimited longevity, among the world's 100 largest economies, 51 were corporations, 49 were countries.[353]
- There are individuals who profess, as a form of dismissal, that links between cancer and environmental damage and contamination are unproven, and further cannot be proved. But isn't it wrong to place innocent, unaware individuals in harms way?
- A good number of the populace is being poisoned daily, some are slightly aware, some have no knowledge of this poisoning whatsoever. Regardless of whether the exact mechanism of this poison can be identified, the data points heavily in the direction that chemicals do indeed cause cancer. Isn't there a duty to inform?
- Is it not a constitutional right to know what freedom now means in this great land? Where there is a desire to know there is a definite duty to inquire. And, until a clear, unbiased answer is obtained, is there not a duty to protect the innocent?[354]
- The history of contamination in the environment is simply the saga of a few publicly held corporations on one hand, and government health officials on the other.[355]

- When we see nations increasingly doing the corporations bidding, becoming fellow degraders and oppressors, we must ask, "How is it that our governments protect living, breathing human beings less and less while protecting corporations more and more? By what authority do our governments trade away to corporations the powers and responsibilities of citizens?" We must work strategically to shift government protection from the corporations to the people.[356]

Motion: The only way to create a safer world is to boycott all toxic products —all products that are genetically altered, all products that contain toxic chemicals, and all products where the safety level for their processing is way below any reasonable standard (factory farms). We can't wait for governments to decide that the burden of proof should **not** be on those potentially at risk. And yet, remember earlier our discussion of the Federal Aviation Administration stopping the flying of all Boeing MD-80 airplanes in the U.S. in January 2000 until each underwent a safety inspection for faulty rear wing stabilizer systems, the suspected cause of an Alaska Air crash killing 88 people. The loss of flying time and income to all U.S. air carriers did not keep the government from acting. Eighty-eight airline deaths can cause an action like this, but 521,000 U.S. cancer deaths a year (in 1992), with an estimated 30,000 deaths attributed to chemical exposure known to be linked to environmental causes,[357] cannot provoke a similar philosophy of erring on the side of caution?

- Sovereignty is in our hands, but the logic is the same: when people running a corporation assume rights and powers which the sovereign have not bestowed, or when they assault the sovereign people, this entity becomes an affront to the body politic. And like a cancer ravaging a human body, such a rebellious corporation must be cut out.[358]
- More than 66% Americans say they want more balance in their lives and want to simplify their lives.[359]

- Between 1990 and 1996, nearly 19% of adult Americans made a voluntary lifestyle change that entailed earning less money.[360]
- There is no right to escape detection. There is no right to commit a perfect crime. The constitution is not at all offended when a guilty man stubs his toe. On the contrary, it is decent to hope he will.
- Nor is it dirty business to use evidence a defendant himself may furnish in the detection stage, as with the culprit who reveals his guilt unwittingly with no intent to shed his inner burden. It is no more unfair to use the evidence he thereby reveals than it is to turn against him the clues at the scene of the crime which a brighter, better informed or more gifted criminal would not have left. It is consonant with good morals and the Constitution, to exploit a criminal's ignorance or stupidity in the detection process [Chief Justice Weintraub, State v. McNight, 52 N.J. 35, 52-53, 24 A. 2nd 240 (1968)].
- The corporate charter is granted by state legislatures. Without a charter a corporation ceases to exist.[361]
- Try to live within the world average income ($1,250 a year) for one month.
- Abolish money and live on barter.

THE POWER OF ONE

Do you often find yourself thinking, what difference can my opinion alone make? "If you think you are too small to make a difference, try sleeping in a closed room with a mosquito."[362] Many of the environmental issues that remain unsolved have more and more become our responsibility as conscientious individuals—air quality and traffic congestion, poor water quality from excess lawn and garden watering and chemical use, runaway energy bills, overflowing landfills. In the search for global sustainability it will be necessary to rethink our values as individual people. In doing so, we will automatically reconsider our roles as consumers and users of resources, homeowners, drivers and commuters.

But to most of us, the problems of pollution and how to prevent global warming, for example, seem to big for the individual to solve. You might be amazed, however, at how even the simplest actions really do count.[363] Think about a day in your life.

*   *   *

Out of bed, on with the warm robe—its not really colder since we turned the thermostat down a degree and the heating bill is looking better. Make a cup of coffee, only heating enough water for one cup of instant saves electricity, water, and time. Jump into the shower instead of taking that bath, using less water and it is also quicker. Brush my teeth…turning the water faucet off between rinses helps save water. Feed the cat, throwing the empty can into the recycling bag to drop at the recycling center on the way to work. Every year the average American discards 1,500 pounds of trash. Making products from recycled material uses 30-55 percent less energy for paper, 33 percent less for glass, and 90 percent less for aluminum.

Quick, check around the house on the way out, turning the thermostat down and turning off all lights. Hear a car horn honk outside. That will be Sally picking the kids up for school as part of our neighborhood car pool. Her daughter has asthma and only one third of the total miles driven by our three-family car pool cuts down on air pollution during rush hour, helping her daughter breath.

After the kids have left I get on my bike and cycle to work. First thing I do when I get to work is turn on my computer—I don't leave it on overnight anymore since I heard about all the electricity it uses. Message from finance on my e-mail about the availability of low interest loans to employees who regularly use public transport in traveling to work. American drivers use 10 billion gallons more gasoline annually than they did in 1989. Coffee break and quick visit to the rest room. Noticed that the firm has installed low flow flush toilets like we have at home (old vintage toilets use 4-6 gallons of water while low-flush models use 1.6 gallons) and is now supplying paper towels made from recycled paper.

Ready to go home. Stop at the market and buy local, seasonal vegetables for dinner, reducing the impact of transport and energy costs. Also, refill my clothes washing powder container from the bulk supplies section of the market. Ready for bed, kids tucked in, all lights off, and just made a reminder note to fix the dripping hot water bath faucet tomorrow. Hot water leaks not only are a waste of water, but are a waste of the energy used to heat that water.

*   *   *

Everything is so connected. The data continually shows that one poison can affect many systems just as one person can incite change in so many. Consider the fact that each American cutting their meat consumption by 10% could free up enough grain to feed the world. Isn't the Power of one amazing? Just because the entities that are doing a large part of the damage are huge and have quite a bit of money, the power of one is still enormous.

USING COMMON SENSE AND DOING GREAT THINGS

Arcata, California (USA) had a big problem in the late 1970s. It had an antiquated sewage treatment system that was dumping raw sewage into Humboldt Bay. The City also did not have the money to share in the cost of building a regional high-tech treatment plant, that probably would spawn all sorts of other growth problems as well. What Arcata did have was a lot of open space, which many large urban communities do not.

Few dream of doing great things with sewage. Robert Gearheart, a sanitation engineer, and George Allen, a fisheries expert, proposed to turn a local Arcata garbage dump into a low-tech treatment plant, wildlife refuge, and salmon-spawning habitat.[364] They set about designing and building the Arcata Marsh and Wildlife Sanctuary which today hundreds of towns in the U.S. and abroad have copied for treating their sewage.

But the marsh holds appeal for more than just sanitation engineers. Just minutes from downtown Arcata, herons, snowy egrets, pelicans, and clapper rails hunt the marsh's chain of ponds. Arcata has become a tourist destination. Every year 150,000 people flock to this town of 17,000 to visit the 154 acre refuge (and sewage treatment plant). Few visitors are aware that the marsh plants are drawing bacteria and other toxins out of the ponds in which they grow. Filter-feeding organisms in the marsh water eat the bacteria that attach to the plants. The ponds serve as a giant filtration system, cleansing the water before it goes into Humboldt Bay. The system needs neither the massive infrastructure nor heavy chemical treatment used to remove bacteria in more traditional sewage plants.

Birders and eco-tourists come to observe the thousands of migrating birds that rest and feed in the Arcata sanctuary every spring and fall. The tourists often don't even realize that the sanctuary's paths lead them through part of Arcata's sewage treatment plant. And the project saved the City considerable money. Its $5.3 million price tag was less than half what they would have needed to share in a regional wastewater treatment project. In addition, because of its tourist attraction, the water treatment/wildlife sanctuary has directly contributed to the region's economy...a win/win situation all the way around.

George Allen was quoted as saying, "it is hard to imagine that two decades ago, the birth of this peaceful place was preceded by battles so bitter they are known in Humboldt County as the 'Wastewater Wars.'" Allen and Gearheart say the marsh battle was really about common sense, about using what is at hand and getting the most out of resources without diminishing their value. Including the eco-tourism component in their vision and project implementation, Arcata significantly added value to their existing resources while solving a very pressing environmental problem.

And, what about trying to grow the perfect lawn? The importance of the perfect lawn is easily understood after considering the circumstances of the U.S. east coast's summer drought of 1999. When asked what people were going to do in light of the water rationing that was implemented for much of the middle-Atlantic region of the United States about things like watering their lawns and car washing rituals, people in New Jersey responded as follows. Well, I guess we will just spray our dying lawns with green paint so that they still look "healthy."

So we have come to a place where our identities teeter on the color of our lawns. Our happiness teeters on the grass presenting as green. It doesn't matter how it got green, just as long as it is green. This detail, this one absurd solution to a drought situation, says it all. Sure, lets just spray our grass green, then have a cookout.

SOCIETAL ACTIONS: THE NEXT INDUSTRIAL REVOLUTION

Most traditional production, such as agriculture or in natural ecosystems, is ecologically dependent, in the sense that it depends directly on certain environmental factors such as sunshine, rainfall, surface water, soil fertility, and/or the availability of genetic materials suitable for seed (reproduction) or medicine. In industrial production, however, environmental factors less directly affect these operations, only really coming into play with regards to serving as resource sources and/or a place for waste disposal in all its forms. There will be fewer incentives to respect ecological constraints in these cases and consequently greater necessity for regulatory intervention of some kind, whether command and control or through the creation of additional economic incentives.

In order to meet these challenges, "we must shift from living our lives to living our lifestyles (consumerism). The world of commerce is the engine for change and can be influenced by re-designing consumerism."[365] Now we can come to a place, a finale so to speak, where the emphasis is on those wonderful, positive, and progressive things that are now happening or hold strong potential to happen in the future. We as a people, and a country, and as citizens of the Earth have very worthy reasons to look at the future with hope. A few years ago, Bill McDonough and Michael Braungart published an article in the Atlantic Monthly magazine that presented their thoughts on a new way of "doing business" for the future.[366] These ideas are so profound and stimulating that to honor them fully is to simply relay them. Therefore, we present excerpts from this 1998 article below.

Recently, some leading industrialists have begun to realize that traditional ways of doing things may not be sustainable over the long term. The 1992 Earth Summit in Rio de Janeiro, led by the Canadian businessman Maurice Strong, recognized those limits. Approximately 30,000 people from around the world, including more than a hundred world leaders and representatives of 167 countries, gathered in Rio de Janeiro (Brazil) to respond to troubling symptoms of environmental decline.

A common idea among many included the fact that a new consciousness or way of thinking about economic and community development may create new ways of "doing" business. Sustainable solutions are likely to be found in the WAY things are done as much as in what is done. Recognizing that what goes around comes around, in terms of coming around again or full circle, citizens can achieve solutions to sustainability by considering the following ways of approaching issues.

- Promote economic development while preserving natural ecosystems.
- Don't ignore the many connections among economic development, natural resource management, and long-term sociologic and ecologic productivity.

- Balance economic development with environmental quality, technological innovation with community stability, and investment in people with investment in infrastructure.
- Develop collaborative partnerships and public consultation processes, across all sectors of society. Everyone involved will ultimately produce a better outcome.

Enter the idea of "eco-efficiency." The hope among industrialists was that eco-efficiency would transform human industry from a system that takes, makes, and wastes into one that integrates economic, environmental, and ethical concerns. Eco-efficiency is now considered by industries across the globe to be the strategy of choice for change.

Plainly put, eco-efficiency aspires to make the old, destructive system less so. But its goals, however admirable, are fatally limited. Produce more with less, "minimize waste," "reduce," and similar dictates advance the notion of a world of limits—one whose carrying capacity is strained by growing populations and exploding production and consumption. Eco-efficiency tells us to restrict industry and curtail growth—to try to limit the creativity and productiveness of humankind.

But, the idea that the natural world is inevitably destroyed by human industry, or that excessive demand for goods and services causes environmental ills, is an over-simplification. Nature—highly industrious, astonishingly productive, and creative, even "wasteful"—is not efficient but effective.

What about nature's method? Consider the cherry tree. It makes thousands of blossoms just so that another tree might germinate, take root, and grow. Who would notice piles of cherry blossoms littering the ground in the spring and think, "How inefficient and wasteful"? The tree's abundance is useful and safe. After falling to the ground, the blossoms that don't produce a new tree return to the soil and become nutrients for the surrounding environment. Every last particle contributes in some way to the health of a thriving ecosystem. "Waste equals food"—the first principle of the Next Industrial Revolution.

COPYING NATURE IN INDUSTRY: AN INTRIGUING IDEA

The cherry tree is just one example of nature's industry, which operates according to cycles of nutrients and metabolisms. This cyclical system is powered by the sun and constantly adapts to local circumstances. Waste that stays waste does not exist.

Human industry, on the other hand, is severely limited. It follows a one-way, linear, cradle-to-grave manufacturing line in which things are created and eventually discarded, usually in an incinerator or a landfill. Unlike the waste from nature's work, the waste from human industry is not "food" at all. In fact, it is often poison. Thus, the two conflicting systems: a pile of cherry blossoms and a heap of toxic junk in a landfill.

According to McDonough, there is an alternative—one that will allow both business and nature to be fecund and productive. This alternative is what he calls "eco-effectiveness." The concept of eco-effectiveness leads to human industry that is regenerative rather than depletive. For a comparison of McDonough's different ideas of industry through the 20th century, comparing the traditional industrial revolution to ideas of eco-efficiency and eco-effectiveness, see Table 2.

The Next Industrial Revolution can be framed as an activity that:
- introduces no hazardous materials into the air, water, or soil;
- measures prosperity by how much natural capital we can accrue in productive ways;
- measures success by how a development process actually enriches the environment;
- measures productivity by how many people are gainfully and meaningfully employed;

TABLE 2. Comparison of characteristics described by the early 20th century Industrial Revolution, the Industrial Revolution with the incorporation of the idea of "eco-efficiency", and the Next Industrial Revolution based upon the concept of eco-effectiveness.367

20th Century Industrial Revolution	1992 Idea of Eco-Efficiency in Industry	Next Industrial Revolution: Eco-Effectiveness
puts billions of pounds of toxic material into the air, water, and soil every year	releases fewer pounds of toxic material into the air, water, and soil every year	introduces no hazardous materials into the air, water, or soil
measures prosperity by activity, not legacy	measures prosperity by less activity	measures prosperity by how much natural capital we can accrue in productive ways
requires thousands of complex regulations to keep people and nature systems from being poisoned too quickly	meets or exceeds the stipulations of thousands of complex regulations that aim to keep people and natural systems from being poisoned too quickly	measures productivity by how many people are gainfully and meaningfully employed
produces materials so dangerous that they will require constant vigilance from future generations	produces fewer dangerous materials that will require constant vigilance from future generations	measures progress by how many buildings have no smokestacks or dangerous effluents
results in gigantic amounts of waste	results in smaller amounts of waste	does not require regulations whose purpose is to stop us from killing ourselves too quickly
puts valuable materials in holes all over the planet, where they can never be retrieved	puts fewer valuable materials in holes all over the planet, where they can never be retrieved	produces nothing that will require future generations to maintain vigilance
erodes the diversity of biological species and cultural practices	standardizes and homogenizes biological species and cultural practices	celebrates the abundance of biological and cultural diversity and solar income

- measures progress by how many buildings have no smoke-stacks or dangerous effluents;
- does not require regulations whose purpose is to stop us from killing ourselves;
- produces nothing that will require future generations to maintain vigilance; and
- celebrates the abundance of biological and cultural diversity and solar income.

This sounds an awful lot like sustainable development. Operationalizing this industrial notion of sustainable development to many seems almost insurmountable. McDonough has talked about a practical approach to acting on the theories of sustainable development in this regard. He states that DESIGN of any project or program is a signal of intention. "Does a design depreciate or enrich people and communities?" Consider for example, "what is the intention in the design of jewish concentration camp?" A mark of a sustainable society is how it treats its least abled citizens.

Likewise, humility in the human circumstance is characterized by our not knowing what to do, but our acknowledging we have to do something different. Consider the idea of cradle to cradle; not cradle to grave. Looking at our legacies instead of our activities. With these intents, McDonough and colleagues developed what they call the "Hanover Principles", which state the following.

- Insist on health and equal rights for all;
- Recognize interdependence;
- Respect relationships between spirit and matter;
- Accept responsibility for the consequences of design;
- Create safe objects of long-term value;
- Eliminate the concept of waste;
- Rely on natural energy flows;
- Understand the limitations of design (nature as a mentor and model…biomimicry); and
- Seek constant improvements by the sharing of knowledge.

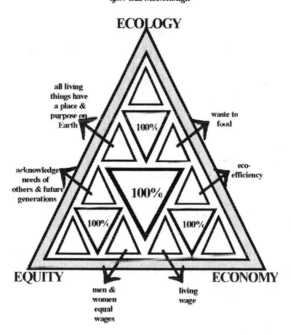

15 The Fractal Ecology Triangle conceptualized by Bill McDonough for explaining the balance that must exist between ecology, equity, and economy in any industrialized society for it to ultimately be sustainable.

For the actual carrying out of these ideas in actions, McDonough has offered a very practical conceptual model that will allow people to actually put into practice (operationalize) the theory of integrating ecology, equity, and economy. This model is referred to as "A Fractal Ecology of Sustainability" (as demonstrated in Figure 15) and offers insight into how you make sustainable decisions around the three points of a triangle, ecology, equity, and economy. In this manner the goals to achieve in taking action can be considered in concert, not in conflict.[368] Any decision is trying to maximize the sustainable outcome (100%) by equally considering both ends of each side of the triangle.

In the context of our earlier discussion of sustainable development's three basic principles, let's examine how these are characterized in an eco-efficient world. **Economy** refers to market viability. Does a product reflect the needs of producers and consumers for affordable products? Safe, intelligent designs should be affordable by and accessible to a wide range of customers, and profitable to the company that makes them, because commerce is the engine of change. **Ecology**, of course, refers to environmental intelligence. Is a material a biological nutrient or a technical nutrient? Does it meet nature's design criteria: waste equals food, respect diversity, and use solar energy? **Equity** refers to social justice. Does the making of new products actually devalue or improve the life of people and communities?

The implications of this operational approach to carrying out sustainable development activities does not assume a perfect world. For example, 90 percent of the time during NASA's Mars Flight, the space craft was off-course. Mid-course adjustments are what made the mission successful.[369] People can do the same in their quests for sustainability, remaining totally flexible and adaptive so that they can modify their course based upon what the progress in their efforts is telling them so far in moving toward their ultimate objective. Plan as much as possible, but always be ready for discovery.

So what does copying nature get us? If someone were to examine the 20th Century Industrial Revolution, it might look like a system of manufacturing as described in column one of Table 2. If we add the idea of eco-efficiency as recognized by industrialists in 1992, our next Industrial Revolution might look like the description in column two of Table 2. But if we want to instead change the whole formula to reflect the idea of eco-effectiveness, the Next Industrial Revolution would look like an industrial system for the next century as described by the listing in column three of Table 2.

READY TO TAKE ACTION?

"Never regret. If it's good, it's wonderful. If it's bad, it's experience."

Victoria Holt

In general today, especially with so much prosperity in places like the U.S., people are turned off by being told what they "need to do" to possibly change things. Change happens much more effectively, with wide-spread results, when people believe they are making decisions because things seem "right" to them personally, not because they are being told by somebody else what to do. Therefore, we have purposely posed all the questions in these pages as an alternative to stating that people **need** to do things. Hopefully we have stimulated your creative juices in thinking about issues from your own particular point of view toward a more collective wisdom, in the context of your day-to-day affairs. Only in this way, from the bottom...up, will sustainability really happen. Legislating it or "needing to do it" won't cut it for the regular global citizen.

Likewise, a large portion of our global population has the notion that environmental protection is an albatross designed to cripple economic and social advancement.[370] A lack of recognition for the connectedness of things has been partially responsible for this view. But the outcome has been adverse environmental impacts, economic failures, social programs that do not work, as well as the many difficulties in achieving world-wide sustainable societies that this perspective creates.

Recent ecologic-economic linked modelling exercises[371] demonstrated that continual learning on a basic, fundamental level, at the grassroots, is crucial in order to achieve sustainability. Human society will never get everything exactly right, but by continual learning we may come close enough to sustain our society and the ecosystems upon which we depend.[372]

The easiest, and probably most effective path is to focus upon the simplicity that really is sustainable development...the parallel, simultaneous consideration of environment, life, and societal well-being. Sustainable communities are all about integrating social, economic, and environmental concerns, rather than addressing each one alone. Afterall, environmentalism means nothing to someone with an empty stomach.[373] Should environmentalists be interested in programs to solve the economic problems of the poor? Absolutely! It is not OK that anyone starves. The goal should be to feed everyone without wrecking the environment.

What may not seem simple, however, is that this also requires considering the large complex issues of population size, climate, energy, resource use, waste management, biodiversity, watershed protection, technology, agriculture, safe water supplies, international security, politics, green building, sustainable cities, smart growth, community/family relations, human values, etc. All these "pieces" are parts of the sustainable society puzzle, because they are the basic ingredients of everyday life. These global concerns about the environment, population increases, foreign policy, climate change are big issues that can easily overwhelm us. It is a natural reaction to simply put these things out of our minds and carry-on with everyday living.

In committing to the goal of sustainable development, however, communities are forcing themselves to keep the whole picture in mind, to try to find the root causes to their problems.[374] Likewise, they are not asking environmentalists to become experts in economics or the larger social issues of poverty, but by encouraging sustainability these disciplines will start working together. That's the splendor of sustainable development: coalitions become the outcome of collaborative efforts among social, economic, and environmental concerns. It forces us to keep an eye on the big picture for the long-term.

In our quickly advancing technology and information age, it is also imperative that people begin to "think outside the traditional box" if they are to fit into the technological society of tomorrow and have the ability to achieve a sustainable future through informed and effective decision-making.[375] In order to find solutions to complex, multifaceted problems related to living sustainably, people today must be exposed to diverse, multiple issue thinking, creative imagining, advanced technology, cross-cultural communication, and environmental ethics and values.[376] This life-long, integrative approach to learning will result in the production of a new society both sensitive to the intrinsic value and inherent worth of the natural environment and responsive to the fact that science and technology should be used for nature's sake and not simply as a means to exploit nature for society's continued use.[377]

A necessary ingredient to moving change in a certain direction, without complete surprise, is having a clear vision of the desired outcome which is also truly shared by all other members of society (organizations and communities).[378] Therefore, the most critical task facing humanity today is the creation of a shared vision of a sustainable and desirable society, one that can provide permanent prosperity within the biophysical constraints of the real world in a way that is fair and equitable to all of humanity, to other species, and to future generations.[379] But how can we really achieve this goal in a world of special interests, inconsistent science, and massive conflicts.

Yankelovich [380] has described the crisis facing modern societies as one of moving from public opinion to public judgement. Public opinion is notoriously fickle and inconsistent on those issues for which the public has not confronted the system-level implications of their opinions. Coming to judgement requires awareness, understanding, and resolution, action, or motion. Sound familiar? A prerequisite for all these steps is bridging the gap between expert knowledge ("culture of technical control") and the public.

But information in the modern world is compartmentalized and controlled by various technical elites who do not communicate with each other.[381] The result is that experts from various fields hold contradictory beliefs and the public holds inconsistent and volatile opinions. Coming to judgement is the process of confronting and resolving these inconsistencies by dissolving the barriers between the mutually exclusive compartments into which information has been put. For example, many people in opinion polls are highly in favor of more effort to protect the environment, but at the same time, they are opposed to any diversion of tax revenues to do so.[382] Coming to judgment is the process of resolving these conflicts and moving to action.

As the authors of this book, if we had a magic wand to address any of the world threats, it would be that henceforth every child would be a wanted and protected child. Because of our children, we have a desire to contribute to a more perfect world for the future. We also want them to have very high expectations, because if they can't imagine it, explain their vision, and have some tools to do their part, sustainability will never happen. Attaining sustainability may well start with merely getting people to ask...What If? If we offer our children poorly designed development and if we accept an ugly, ignored environment, how will our children have any idea that it could be different? When the majority of community members make sustainability a part of everyday decisions while shopping, building, recreating, working, and voting, then these communities will achieve sustainability.

Epilogue

The Wall (as in Berlin Wall) that once held so many hostage in Eastern Europe, the Wall that was made of cement, block and mortar, finally did fall. And as this Wall fell, as this physical blockade crumbled, as the last identifiable remnants of this monstrous prison slowly disappeared, hope was genuinely felt. This hope was felt by those on both sides of the Wall: and many prayed for freedom. Yet this hope somehow got pushed aside in the chaos of change. And once again people are held captive by something much more subtle and sublime than their former isolating barricade. They are now held in check by a mentality that is enforced by a foreign mafia, self-appointed street patrols, and a lost, helpless, leaderless army. Somehow their oppression is recreated. At some level, to some degree, their thoughts deemed their existence to be eternally one of some form of captivity. Not a conscious process, yet a very powerful process none-the-less.

Thought is a powerful thing. Thought to word brings more power. Word to voice more, and voice to motion (action) changes patterns, lives, and worlds. Within the labyrinth of life it is in thought that all begins. Thought is ignited from two places: from old conditioned comfort levels or from a remembering of the soul. The phrase "old habits die hard" should not be taken lightly here. For it is in thought that these old habits regenerate. And it is these old thought patterns and comfort levels in which denial lives—a denial that kills. Denial kills at many levels and to all degrees as it ultimately masks itself in a blaming stance that keeps one stuck in a specific existence—stuck in that long hard day; stuck in that state of unconscious movement; stuck in a place of absurd redundancy and regret. To look back on one's life and wish to have lived differently due to personal loss is a sad, sad thing. But to look back on one's life and to wish to have lived differently for the very existence of one's own flesh and blood is a grief that kills with a dull, repetitious jab to the heart.

We as citizens of countries are squarely rooted in many, many addictions. We as individuals of the most powerful countries of this age have given our power away to thirty minute sit-coms. And, we as a people of a most wondrous and diverse lands have put blinders on that allow this denial not only breath and life, but allow our addictions, and the denial therein, to own us.

We are innately good. Human nature is loving and compassionate. Humans and nature share a name (human-nature) because our existence is so intricately interwoven on this Earth. If we destroy the air, land, water, and life around us, we destroy ourselves. If we deny this destruction, we deny ourselves, our children, and our children's children all the beauty destiny holds. It is time to bridge the gap between our hearts, our minds, and our souls. Somewhere in our beings we all know that it is from the Earth that we came, and it is in this Earth that we shall rest. And it is in this remembering that we will learn, again, to dance our dance of life without harming the land that we belong so eternally to.

*   *   *

I have come to terms with the future.
From this day onward I will walk easy on the earth.
Plant trees. Kill no living things.
Live in harmony with all creatures.
I will restore the earth where I am.
Use no more of its resources than I need.
And listen; listen to what it is telling me.

M.J. Slim Hooey

About the Authors

Dr. Warren Flint obtained a Ph.D. in Ecology from the University of California, Davis in 1975. He has 25 years of university research and teaching experience in environmental science, marine biology, and ecology. Over the last decade Dr. Flint has become a sustainable development specialist, working and writing on the topic to more clearly understand and articulate a sound philosophy to guide healthy economic growth while also protecting/rehabilitating environmental systems and enhancing social equity and well-being. He now serves as principal for Five E's Unlimited, a firm that provides international leadership in developing research, management, policy, and education strategies to solve problems of environmental degradation, economic decline, and community disintegration as defined by governments, corporations, academic institutions, communities, and non-governmental organizations. He is internationally recognized for initiating and directing interdisciplinary research targeting ecological and natural resource issues and has authored 3 books and more than 60 peer-review journal articles on diverse environmental science subjects. Dr. Flint also regularly gets involved in community capacity building by advancing sustainable development practices in community revitalization, guiding environmental fact-finding, facilitating conflict resolution, promoting consensus-building in participatory decision-making processes, and assisting in community-based natural resource and comprehensive watershed planning.

Willow Lisa Houser holds a Master's Degree in addictions rehabilitation from Wright State University (1991). She runs her own business,

Addiction Research Center. In 1993 she directed an outreach program entitled "Saturday's Child" that won the Best Prevention Program Award two years in a row from the Governor of Ohio (USA). She has worked consistently in the field of addictions and through public speaking, Willow is able to communicate with hard-to-reach populations, simplifying information and teaching people how to accommodate new information and behaviors into daily life. Ms. Houser has held a great interest in contemporary and cutting-edge environmental concerns for the past 7 years. She supports many environmental causes as well as adheres to harmless living to whatever degree is feasible. In 2000 she founded Beyond Creation, a non-profit organization to advocate awareness of renewable energy sources, dominantly wind energy, and promote sustainable community development by expanding environmental education activities with children. Beyond Creation advocates pro-environmental issues with the use of a solar-powered playhouse she and her children built, which is equipped with Enviroscapes that demonstrate various effects waste has on nature as well as different sustainable-based solutions. Her other publications include "An Era of Addiction: The Evolution of Dependency" which discusses Cultural Addiction.

Appendix I

Pertinent Internet Web Sites on Sustainable Development

Below is a listing of World Wide Web (www) sites that offer information on different aspects of sustainable development. We do not necessarily endorse any one of these sites just because of their appearance here. This list is intended only as a reference for people wanting to explore different ideas and issues involving sustainable development.

- American Forests - Mapping Urban Ecology: Benefits of trees, soils, and other natural resources
  http://www.americanforests.org/ufc/cgreen/cgad.html
- Applying Sustainable Development
  http://www.applysd.co.uk
- Aspen Institute
  http://www.aspeninst.org/rural/updates
- Association of University Leaders for a Sustainable Future
  http://www.ulsf.org
- Austin Sustainable Building Coalition
  http://www.greenbuilder.com/general/articles/sbc.html
- Best Foot Forward
  http://www.bestfootforward.com

- The Center for Environment & Business on the Web
  http://sustainablebusiness.com
- The Center for Innovation in Corporate Responsibility
  http://www.triplebottomline.com
- Center for Respect of Life and Environment
  http://www.crle.org
- Center of Excellence for Sustainable Development: U.S.
  Department of Energy
  http://www.sustainable.doe.gov
- Centre for Development & Environment (CDE)
  http://www.cde.unibe.ch/index.html
- Centre for Sustainable Design
  http://www.cfsd.org.uk
- Centre for a Sustainable Economy
  http://www.sustainableeconomy.org
- Climate Solutions: Practical Solutions to Global Warming
  http://www.climatesolutions.org
- Coalition for Environmentally Responsible Economies (CERES)
  http://www.ceres.org
- Coalition for Environmentally Responsible Economies: Global
  Reporting Initiative (GRI)
  http://www.globalreporting.org/AboutGRI.htm
- Communications for a Sustainable Future
  http://csf.colorado.edu
- Communities by Choice
  http://www.communities-by-choice.org
- Community Development Society Sustainability Bibliography
  http://comm-dev.org/sections/practice/biblio.htm
- Creative Change Educational Solutions
  http://www.creativechange.net
- Earth Charter USA
  http://www.earthcharterusa.org
- The Earth Systems Organization
  http://earthsystems.org

- EcoIQ Magazine Online
  http://www.ecoiq.com/magazine
- Ecologic Institute International, Inc.
  http://eiii.org
- EcoMall - Source of environmentally friendly companies and products
  http://www.ecomall.com
- EcoSTEPS: Australian triple bottom line sustainability
  http://www.ecosteps.com.au/body1.htm
- Ecotopia
  http://ecotopia.com/index.asp
- Ecovillage Network of the Americas
  http://ena.ecovillage.org
- Education for Sustainability
  http://csf.concord.org/esf
- The Enertia Kit - Green Builders
  http://enertia.com/build_it.htm
- EnviroLink Network
  http://www.envirolink.org/envirohome.html
- The Environmental Atlas for Researching Environmental Policy
  World-wide
  http://www.rri.org/envatlas/index.html
- Environmental Organization's Web Directory
  http://www.webdirectory.com
- Environmental Resources
  http://www.geocities.com/rdkii
- Environmental Taxation - How to design/implement environmental
  taxes
  http://www.greentaxes.org/intro.asp
- European Foundation for the Improvement of Living and Working
  Conditions
  http://susdev.eurofound.ie
- The European Sustainable Cities & Towns Campaign
  http://www.sustainable-cities.org

- E-VOLVE: for those who wish to respond to change
  http://www.e-volve.org.uk
- Facing the Future: People and the Planet
  http://www.facingthefuture.org
- Five E's Unlimited - Sustainable Development Specialists
  http://www.eeeee.net
- Florida Sustainable Communities Center
  http://sustainable.state.fl.us
- Fostering Sustainable Behavior
  http://www.cbsm.com
- Foundation for Sustainable Development
  http://www.unc.edu/~arobb/fsd
- Future Generations
  http://www.future.org
- Gallup's Earth Day 2000 Poll
  http://www.gallup.com/poll/releases/pr000418.asp
- Global Action Plan for the Earth
  http://www.globalactionplan.org
- Global Green USA
  http://www.globalgreen.org
- Global Resource Bank
  http://www.grb.net
- Global System for Sustainable Development
  http://gssd.mit.edu
- Global Warming: Early Warning Signs
  http://www.climatehotmap.org
- Green Communities Program - U.S. EPA, Region III
  http://www.epa.gov/region3/greenkit
- Green Mountain Institute
  http://www.gmied.org
- Hart Environmental Data
  http://www.subjectmatters.com/indicators
- Heartland Center for Leadership Development
  http://web.4w.com/heartlandcenter

- Human Sustainability: The Example of Kerala, India
  http://www.jadski.com/kerala
- International Institute on Sustainable Development
  http://iisd.ca
- International Rural Development - The World Bank
  http://www.worldbank.org/devforum
- Issac Walton League's Sustainability Education Project
  http://www.iwla.org/sep/
- Journal on Sustainability in Higher Education
  http://www.mcb.co.uk/ijshe.htm
- Loka Institute - Community Research Center Programs
  http://www.loka.org
- Marshall Erdman Academy of Sustainable Design
  http://erdman.com/'academy/index.html
- Miscellaneous Rural Sustainable Development Links
  http://www.links2go.com/channel/Sustainable_Development
- Mountain Association for Community Economic Development -
  MACED
  http://www.maced.org
- The National Centre for Sustainability Society (Canada)
  http://www.islandnet.com/~ncfs/ncfs/
- The Natural Step - US
  http://www.naturalstep.org
- Organization for Economic Cooperation & Development - OECD
  http://www.oecd.org
- Overcoming Consumerism
  http://www.hooked.net/users/verdant/index.htm
- Population & Environment Linkages Service
  http://popenvironment.org
- Population, Health, and Environmental Issues
  http://www.popplanet.org
- Project on Sustainability Centers
  http://www.projekte.org/sustainabilitycentres

- Rachel's Environment and Health Weekly
  http://www.rachel.org
- Resilient Communities
  http://www.resilientcommunities.org
- Rocky Mountain Institute
  http://www.rmi.org
- Rural Sustainable Development Indicators
  http://www.hq.nasa.gov/iwgsdi
- Second Nature, Inc.
  http://www.secondnature.org
- SEED - A feminine way toward values-led sustainable enterprise
  http://www.seedfusion.com
- Selby Resource Collection on Sustainable Development
  http://www.olt.qut.edu.au/int/selby/background/Default.cfm?Res=5
- Sigma XI Sustainable Development
  http://csf.colorado.edu/casx
- Smart Growth
  http://www.smartgrowth.org
- Southface Energy Institute - Green Building Program
  http://www.southface.org
- Sustainable America Town Meeting
  http://www.sustainable-usa.org
- Sustainable Business Network
  http://sbn.envirolink.org
- Sustainable Business Resources
  http://www.sustainablebusiness.com
- Sustainability Centre in Germany
  http://www.econtur.de
- Sustainable Chattanooga
  http://www.chattanooga.net/SUSTAIN/index.html
- Sustainable Communities
  http://www.sustainable.org
- Sustainable Communities - U.S. Environmental Protection Agency
  http://www.epa.gov/ecocommunity

- Sustainable Development
  http://www.lib.kth.se/~lg/sustain.htm
- Sustainable Development: Best Starting Points
  http://www.ulb.ac.be/ceese/meta/sustain.html
- Sustainable Development Bibliography
  http://www.comp-dev.org/sections/practice/Bibliography%20-
  %20Sustainability-CDS.htm
- Sustainable Development Dimensions
  http://www.fao.org/sd
  Sustainable Development (SD) Gateway
  http://sdgateway.net
- Sustainable Development Information
  http://www.sdinfo.gc.ca
- The Sustainable Development Institute
  http://www.susdev.org
- Sustainable Development (SD) Online
  http://susdev.eurofound.ie
- Sustainable Development Solutions
  http://www.SustainableDevelopmentSolutions.com
- Sustainable Development - U.S. EPA, Region III
  http://www.epa.gov/Region3/sdwork/index.htm
- Sustainable Futures
  http://www.sustainablefutures.org
- Sustainability Debate - Inviting Debate
  http://www.cyberus.ca/choose.sustain
- Sustainable Seattle
  http://www.scn.org/sustainable/susthome.html
- Terrain: An On-line Journal of the Built & Natural Environments
  http://www.terrain.org
- Thoreau Center for Sustainability
  http://www.thoreau.org
- Toxics Use Reduction Institute
  http://www.turi.org

- United Nations Sustainable Development Web Site
  http://www.un.org/esa/sustdev
- Urban Options - Sustainable Lansing (MI)
  http://urbanoptions.org
- Village Habitat Design - Architectural & Conservation Planning
  http://www.VillageHabitat.com
- The World Resources Institute
  http://www.wri.org
- The World Watch Institute
  http://www.worldwatch.org
- The Yellow Mountain Institute for Sustainable Living
  http://monticello.avenue.gen.va.us/Community/Environ/YellowMtn

Bibliography

RELATED READINGS:

Barker, R. 1997. And The Waters Turned To Blood. Simon & Schuster, New York, NY. 332 pp.

Bell, A. and W. Streiber. 1998. The Coming Global Superstorm. Pocket Books, Simon & Schuster, New York, NY. 255 pp.

Berlin, S. 1997. The Ways We Live: Exploring Community. New Society Publishers, Gabriola Island, British Columbia. 170 pp.

Bernard, T. & J. Young. 1997. The Ecology of Hope: Communities Collaborate for Sustainability. New Society Publishers, Gabriola Island, British Columbia. 233 pp.

Bossel, H. 1999. Indicators for Sustainable Development: Theory, Method, Applications. International Institute for Sustainable Development, Winnipeg, Manitoba. 124 pp.

Brown, L., et al.. 1999. State of the World 1999. The Worldwatch Institute, Washington, DC. 259 pp.

Brown, L.R., M. Renner, and B. Halweil. 1999. Vital Signs 1999: The Environmental Trends that are Shaping Our Future. Worldwatch Institute, Washington, DC and W.W. Norten & Comapany, New York, NY. 276 pp.

Brown, L.R., et.al. 2000. State of the World 2000. The Worldwatch Institute, Washington, DC and W.W. Norten & Company, Inc., New York, NY. 276 pp.

Camacho, D.E. 1998. Environmental Injustices: Political Struggles Race, Class and the Environmental. Duke University Press, Durham, NC. 232 pp.

Chomsky, N. 1997. Media Control the Spectacular Achievements of Propaganda. Seven Stories Press, New York, NY. 58 pp.

Costanza, R., J. Cumberland, H. Daly, R. Goodland, & R. Norgaard. 1997. An Introduction to Ecological Economics. St. Lucie Press, Boca Raton, FL. 275 pp.

Edwards, M. and H. David (eds). 1996. Beyond the Magic Bullet: NGO Performance and Accountability in the Post-Cold War World. Kumerian Press, West Hartford, CT. 283 pp.

Flavin, C. and L. Nicholas. 1994 Power Surge: Guide to the Coming Energy Revolution. Worldwatch Institute, Little, Brown and Company, Boston, MA. 351 pp.

Gibbs, Lois Marie. 1995. Dying Form Dioxin. South End Press, Boston, MA. 361 pp.

Gore, A. 1992. Earth In The Balance: Ecology and the Human Spirit. Houghton Mifflin Co., New York, NY. 407 pp.

Gotlieb, Y. 1996. Development, Environment, and Global Dysfunction: Toward Sustainable Recovery. St. Lucie Press, Delray Beach, FL. 188 pp.

Greer, J. and K. Bruno. 1992. Greenwash: The Reality Behind
Corporate Environmentalism. The Apex Press, New York, NY. 258 pp.

Harrison, P. 1993. Inside the Third World. Penguin Books, London,
England. 529 pp.

Hartmann, T. 1998. The Last Hours of Ancient Sunlight. Mythical
Books, Northfield, VT. 299 pp.

Hawken, P. 1993. The Ecology of Commerce: A Declaration of
Sustainability. HarperCollins Publ., New York, NY.

Hawken. P., et al. 1999. Natural Capitalism. Little, Brown and
Company, Boston, MA. 396 pp.

Isbister, J. 1998. Promises Not Kept: The Betrayal of Social Change
in the Third World. Kumerian Press, West Hartford, CT. 272 pp.

Korten, D.C. 1996. When Corporations Rule the World. Kumerian
Press, West Hartford, CT. 374 pp.

Korten, D.C. 1997. Globalizing Civil Society: Reclaiming Our Right
to Power. The Pen Media Pamplet Series by Seven Stories Books,
New York, NY. 78 pp.

Korten, D.C. 1999. The Post Corporate World. Berrett-Koehler
Publishers, Inc., and Kumerian Press Inc., West Hartford, CT. 318 pp.

Mander, J. and E. Goldsmith. 1996. The Case Against the Global
Economy. Sierra Club Books, San Francisco, CA. 549 pp.

Maser, C. 1996. Resolving Environmental Conflict: Towards
Sustainable Community Development. St. Lucie Press, Delray Beach,
FL. 200 pp.

Maser, C. 1997. Sustainable Community Development: Principles and Concepts. St. Lucie Press, Delray Beach, FL. 257 pp.

McChesney, R. 1997. Corporate Media and the Threat to Democracy. The Pen Media Pamplet Series by Seven Stories Books, New York, NY. 79 pp.

McKibben, B. 1995. Hope, Human and Wild. Hungry Mind, Little Brown and Company, St Paul, MN. 237 pp.

Meadows, D.H., D.L. Meadows, & J. Randers. 1992. Beyond the Limits: Confronting Global Collapse - Envisioning a Sustainable Future. Chelsea Green Publishing Co., Post Mills, VT. 300 pp.

Morris, J.A. 1994. Not In My Backyard: The Handbook. Silvercat Publications, San Diego, CA. 305pp.

Muschett, F.D. (ed.). 1997. Principles of Sustainable Development. St. Lucie Press, Delray Beach, FL. 176 pp.

Nader, R., et. al. 1993. The Case Against Free Trade. Earth Island Press, Berkeley, CA. 230 pp.

Nattrass, B. and M. Altomare. 2001. The Natural Step for Business: Wealth, Ecology and the Evolutionary Corporation. New Society Publishers, Gabriola Island, British Columbia. 240 pp.

The President's Council on Sustainable Development. 1999. Towards a Sustainable America: Advancing Prosperity, Opportunity, and a Healthy Environment for the 21st Century. U.S. Government Printing Office, Washington, DC. 158 pp.

Prugh, T., R. Costanza, J.H. Cumberland, H. Daly, R. Goodland, and R.B. Norgaard. 1995. Natural Capital and Human Economic

Survival. International Society of Economics and Ecology (ISEE) Press, Solomons, MD.

Renner, M. 1996. Fighting for Survival. W.W. Norten & Comapany, New York, NY. 239 pp.

Ridgeway, J. and C. Jeffery. 1998. A Pocket Guide to Environmental Bad Guys. Thunder's Mouth Press, New York, NY. 178 pp.

Robbins, J. 1987. Diet for a New America. H.J. Kramer, Tiburon, CA. 423pp.

Roseland, M. 1998. Toward Sustainable Communities. New Society Publishers, Gabriola Island, British Columbia. 241 pp.

Ryan, J., C. Durning, and A. Thein. 1997. Stuff: The Secret Lives of Everyday Things. Northwest Environment Watch, Seattle, WA. 86 pp. Schaef-Wilson, A. 1998. When Society Becomes An Addict. HarperCollins, New York, NY. 152 pp.

Schmidheiny, S. 1992. Changing Course. The MIT Press, Cambridge, MA. 374 pp.

Shaumatoff, A. 1990. The World Is Burning: Murder in the Rain Forest. Little, Brown and Co., Boston, MA. 377 pp.

Sitarz, D. 1998. Sustainable America: America's Environment, Economy, and Society in the 21st Century. Earthpress, Carbondale, IL. 312 pp.

Steingraber, S. 1998. Living Downstream. Vintage Books, New York, NY. 373 pp.

Theobald, R. 1995. Reworking Success New Communities at the Millennium. New Society Publishers, Gabriola Island BC, Canada. 119 pp.

Thomas, W.L.. 2000. The Age of Addiction. (in preparation). 150pp

Tokar, B. 1997. Earth For Sale: Redeeming Ecology in the Age of Corporate Greenwash. South End Press, Boston, MA. 269 pp.

Wackernagel, M. & W. Rees. 1996. Our Ecological Footprint: Reducing Human Impact on the Earth. New Society Publishers, Gabriola Island, British Columbia. 160 pp.

Endnotes

1. Jacobs, J. 2000. The Nature of Economies. The Modern Library, New York, NY. pg. 138.

2. Jacobs, J. 2000. The Nature of Economies. The Modern Library, New York, NY. pg. 145.

3. Edie Weekly Summaries. 2000. Prince Charles says sustainable development requires new world view. 5/19/2000. [online] URL: http://www.edie.net/news/Archive/2741.html.

4. Edie Weekly Summaries. 2000. Prince Charles says sustainable development requires new world view. 5/19/2000. [online] URL: http://www.edie.net/news/Archive/2741.html.

5. Bartlett, A.A. 1990. Reflections on sustainability, population, growth, and the environment - Revisited. Renewable Res. J. 15(4): 6-23.

6. Roodman, D.M. personal communication (1999). The Worldwatch Institute, Washington, DC. (e-mail: drood@worldwatch.org).

7. Hartmann, T. 1997. Last Hours of Ancient Sunlight. Mythical Books, Northfield, VT. pg. 46. Environmental News Network (ENN), 8/5/99. Humans altering Earth for the Worst. [online] URL: http://www.enn.com/enn-news-archive/1999/08/080599/deadzone_4798.asp

8. Orr, D.W. 1994. Earth in Mind: On Education, Environment, and the Human Prospect. Island Press, Washington, DC. Orr. D.W. 1992. Ecological Literacy: Education and the Transition to a Postmodern World. State University of New Press, Albany, NY. pg. 24.

9.   Roodman, D.M. personal communication (1999). The
        Worldwatch Institute, Washington, DC. (e-mail: drood@world-
        watch.org).
10.   Clark, R. (ed.) 1999. Overview GEO-2000: Global
        Environment Outlook. United Nations Environment
        Programme, Nairobi, Kenya. pg. 4-5.
11.   Pimentel. D. 1999. How Many Americans Can The Earth
        Support? Population Press 5(3): 1-2.
12.   Wackernagel. M. and W. Rees. 1996. Our Ecological
        Footprint. New Society Publ., Gabriola Island, BC, Canada.
        pg. 54.
13.   Environment News Service (ENS), 8/24/99. Canada, U.S.
        Consider Great Lakes Water Export Ban. [online] URL:
        http://ens.lycos.com/ens/aug99/1999L-08-24-06.html
14.   Maser, C. 1997. Sustainable Community Development:
        Principles and Concepts. St. Lucie Press, Delray Beach, FL.
        pg. 43
15.   Booth, E.M. 1996. Starting with Behavior: A Participatory
        Process for Selecting Target Behaviors in Environmental
        Programs. GreenCOM Project, Academy for Education
        Development, Washington, DC. 96 p.
16.   Bernard, T. and J. Young. 1997. The Ecology of Hope:
        Communities Collaborating for Sustainability. New Society
        Publ., Gabriola Island, BC, Canada. pg. 27.
17.   Meadow, D.H., D.L. Meadows, J. Randers, and W.W. Behrens.
        1972. Limits to Growth: A Report for the Club of Rome's
        Project on the Predicament of Mankind. Universe Books,
        New York, NY.
18.   Bartlett, A.A. 1994. Reflections on sustainability, population
        growth, and the environment. Population and Environment,
        16(1): 5-35.

19.   The World Commission on Environment and Development
      (WCED). 1987. Our Common Future. Oxford University
      Press, New York, NY. pg. 43.
20.   Daly, H.E. 1996. Beyond Growth. Beacon Press, Boston,
      MA. pg. 48.
21.   Bartlett, A.A. 1999. Colorado's Population Problem.
      Population Press, 5(6): 8-9
22.   Maser, C. 1996. Resolving Environmental Conflict: Towards
      Sustainable Community Development. St. Lucie Press, Delray
      Beach, FL. pg. 163.
23.   Bernard, T. and J. Young. 1997. The Ecology of Hope:
      Communities Collaborating for Sustainability. New Society
      Publ., Gabriola Island, BC, Canada. pg. 190
24.   Bernard, T. and J. Young. 1997. The Ecology of Hope:
      Communities Collaborating for Sustainability. New Society
      Publ., Gabriola Island, BC, Canada. pg. 184.
25.   Giuliano, J.A. 1998. The Well Will Run Dry. Environment
      News Service (ENS), 11/14/98, Healing Our World - Weekly
      Comment. [online] URL:
      http://ens.lycos.com/ens/nov98/1998L-11-14g.html
26.   Muschett, F.D. 1997. An integrated approach to sustainable
      development, pg. 7-9. *In*: Muschett, F.D. (ed.), Principles of
      Sustainable Development, St. Lucie Press, Delray Beach, FL.
      176 pp.
27.   Bartlett, A.A. 1998. Reflections on sustainability, population
      growth, and the environment. Renewable Resources Journal
      15(4): 6-23.
28.   Wackernagel. M. and W. Rees. 1996. Our Ecological
      Footprint. New Society Publ., Gabriola Island, BC, Canada.
      pg. 115.
29.   Flint, R.W. and M.J.E. Danner. 2001. THE NEXUS OF
      SUSTAINABILITY & SOCIAL EQUITY: Virginia's Eastern
      Shore (USA) as a Local Example of Global Issues. Int.
      Journal of Economic Development (in press).

30.   Warren, J.L. 1997. How Do We Know What is Sustainable?
       In: F.D. Muschett (ed.), Principles of Sustainable
       Development. St. Lucie Press, Delray Beach, FL. pg. 131-
       133.
31.   Korten, D.C. 1995. When Corporations Rule The World.
       Kumarian Press, West Hartford, CT. pg. 49.
32.   Bruno, K., J. Konliner, and C. Brotsky. 1999. Greenhouse
       gangsters vs. climate justice. TRAC/Trides Center. [online]
       URL: http://www.corpwatch.org/climate.
33.   Wackernagel. M. and W. Rees. 1996. Our Ecological
       Footprint. New Society Publ., Gabriola Island, BC, Canada.
       pg. 54. Korten, D.C. 1995. When Corporations Rule The
       World. Kumarian Press, West Hartford, CT. pg. 107.
34.   Giuliano, J.A. 2000. Its Worse Than You Think. Healing Our
       World: Weekly Comment, Environment News Service (ENS),
       2/14/00. [online] URL:
       http://ens.lycos.com/ens/feb2000/2000L-02-14g.html.
       Giuliano, J.A. 1999. Y6B - Its Really Coming. Environment
       News Service (ENS) 6/21/99, Healing Our World - Weekly
       Comment. [online] URL:
       http://ens.lycos.com/ens/jun99/1999L-06-21g.html.
35.   Brown, L. 1999. State of the World: A Worldwatch Institute
       Report on Progress Toward a Sustainable Society. W.W.
       Norton and Co., New York, NY.
36.   Wackernagel. M. and W. Rees. 1996. Our Ecological
       Footprint. New Society Publ., Gabriola Island, BC, Canada.
       pg. 51.
37.   Wackernagel. M. and W. Rees. 1996. Our Ecological
       Footprint. New Society Publ., Gabriola Island, BC, Canada.
       pg. 7.
38.   Ryan, J.C. 1997. Stuff - The Secret Lives of Everyday Things.
       Northwest Environmental Watch, Seattle, WA. pg. 57.

39. Wackernagel. M. and W. Rees. 1996. Our Ecological Footprint. New Society Publ., Gabriola Island, BC, Canada. pg. 3.

40. Ryan, J.C. 1997. Stuff - The Secret Lives of Everyday Things. Northwest Environmental Watch, Seattle, WA. pg. 5.

41. Production estimate for 1998 based upon the AAMA, World Motor Vehicle Facts and Figures, 1998. Washington, DC.

42. Environment News Service (ENS), 7/20/99. Growing population faces shrinking water supply. [online] URL: http://ens.lycos.com/ens/jul99/1999L-07-20-01.html.

43. Ryan, J.C. 1997. Stuff - The Secret Lives of Everyday Things. Northwest Environmental Watch, Seattle, WA. pg. 55.

44. World Commission on Forests and Sustainable Development. 1999. Blueprint to Solve Global Forest Crisis. Washington, DC. [online] URL: http://www.edie.net/news/Archive/1063.html.

45. Maser, C. 1997. Sustainable Community Development: Principles and Concepts. St. Lucie Press, Delray Beach, FL. pg. 162.

46. Lazaroff, C. 1999. Development devouring three million U.S. acres a year. Environment News Service (ENS), 12/7/99. [online] URL: http://ens.lycos.com/ens/1999L-12-07-06.html.

47. Cox, P.A. 2000. Will tribal knowledge survive the millennium? Science 287: 44-45.

48. Hartmann, T. 1997. Last Hours of Ancient Sunlight. Mythical Books, Northfield, VT. pg. 76-80.

49. Thomas, W.L. 2000. The Age of Addiction. (in preparation for publication).

50. Hartmann, T. 1997. Last Hours of Ancient Sunlight. Mythical Books, Northfield, VT. pg. 77-79.

51. Daily, G.C. 1999. Developing a scientific basis for managing Earth's life support systems. Conservation Ecology 3(2): 14. [online] URL: http://www.consecol.org/vol3/iss2/art14.

52.   Kirch, P.V. 1977. Microcosmic histories: Island perspectives on "global" change. American Anthropologist 99: 30-42.

53.   Daily, G.C. 1999. Developing a scientific basis for managing Earth's life support systems. Conservation Ecology 3(2): 14. [online] URL: http://www.consecol.org/vol3/iss2/art14.

54.   Daily, G.C. 1999. Developing a scientific basis for managing Earth's life support systems. Conservation Ecology 3(2): 14. [online] URL: http://www.consecol.org/vol3/iss2/art14.

55.   Hartmann, T. 1997. Last Hours of Ancient Sunlight. Mythical Books, Northfield, VT. pg. 77-79.

56.   Giuliano, J.A. 1999. Y6B - Its Really Coming. Environment News Service (ENS) 6/21/99, Healing Our World - Weekly Comment. [online] URL: http://ens.lycos.com/ens/jun99/1999L-06-21g.html

57.   Hempel, M. 1999. A World of Six Billion. Population Press 5(6): 3.

58.   Lazaroff, C. 2000. Growing Population Faces Diminishing Resources. Environment News Service (ENS) 1/18/00. [online] URL: http://ens.lycos.com/ens/jan2000/2000L-01-18-06.html

59.   Robbins, J. 1987. Diet For A New America. H.J. Kramer, Inc., Tiburon, CA. pg. 357.

60.   Lazaroff, C. 1999. Development devouring three million U.S. acres a year. Environment News Service (ENS), 12/7/99. [online] URL: http:ens.lycos.com/ens/dec99/1999L-12-07-06.html.

61.   Edie Weekly Summaries, 9/24/99. Populations outrunning water supply as world hits 6 billion. [online] URL: http://www.edie.net/news/Archive/1713.html. Environment News Service (ENS), 8/30/99. U.S. Drought Signals Worldwide Water Shortage. [online] URL: http://ens.lycos.com/ens/aug99/1999L-08-30-06.html

62.   Giuliano, J.A. 1998. The Well Will Run Dry. Environment News Service (ENS), 11/14/98, Healing Our World - Weekly

Comment. [online] URL:
http://ens.lycos.com/ens/nov98/1998L-11-14g.html

63.   Hartmann, T. 1997. Last Hours of Ancient Sunlight.
        Mythical Books, Northfield, VT. pg. 13.

64.   Wackernagel, M. and W.E. Rees. 1996. Our Ecological
        Footprint: Reducing Human Impact on the Earth. New
        Society Publ., Gabriola Island, B.C., Canada. pg. 149.

65.   Maser, C. 1997. Sustainable Community Development:
        Principles and Concepts. St. Lucie Press, Delray Beach, FL.
        pg. 170.

66.   Schowengerdt, S. 1999. EPA Taken To The Task Over Pollution
        Data. Environmental News Network (ENN) 10/29/99 [online]
        URL: http://www.enn.com/news/enn-stories/1999/10/102999/chi-
        rality_6863.asp. Environment News Service (ENS) 8/6/99.
        Ignorance Is Not Bliss About Hormone Changing Chemicals.
        [online] URL: http://ens.lycos.com/ens/aug99/1999L-08-06-
        03.html.

67.   Colborn, T. and M.J. Smolen. 1996. Epidemiological analysis
        of persistent organochlorine contaminants in cetaceans. Rev.
        Environ. Contam. Toxicol. 146: 91-172.

68.   Colborn, T., D. Dumanoski, and J.P. Meyers. 1996. Our
        Stolen Future. Dutton Publ., New York, NY.

69.   Gaither, C. 1999. The Low Side to High Tech. Environment
        News Service (ENS) 10/11/99. [online] URL:
        http://ens.lycos.com/ens/oct99/1999L-10-11-03.html.

70.   Ritter, L., K.R. Solomon, J. Forget, M. Stemeroff, and C.
        O'Leary. 1995. A Review of Selected Persistent Organic
        Pollutants. Int. Prog. on Chemical Safety, World Health
        Organization, New York, NY.

71.   Environment News Service (ENS) 2/19/99. PCBs in Texas
        Dolphins a Warning to Humans. [online] URL:
        http://ens.lycos.com/ens/fed99/1999L-02-19-03.html.

72.   Auman, H.J., J.P. Ludwig, C.L. Summer, D.A. Verbrugge, K.L.
        Froese, T. Colborn, and J.P. Giesy. 1997. PCBs, DDE, DDT,

and TCDD-EQ in two species of albatross on Sand Island, Midway Atoll, North Pacific Ocean. Environ. Toxicol. Chem. 16(3): 498-504.

73. Auman, H.J., J.P. Ludwig, C.L. Summer, D.A. Verbrugge, K.L. Froese, T. Golborn, and J.P. Giesy. 1997. PCBs, DDE, DDT, and TCDD-EQ in two species of albatross on Sand Island, Midway Atoll, North Pacific Ocean. Environ. Toxicol. Chem. 16(3): 498-504.

74. Svensson, B., T. Hallberg, A. Nilsson, B. Akesson, A. Schutz, and L. Hagmar. Immunological competence and liver function in subjects consuming fish with organochlorine contaminants. In: Fiedler, H., H. Frank, O. Hutzinger, W. Parxefall, A. Riss, and S. Safe (eds), pg. 175-178, DIOXIN '93: 13th Int. Symp. on Chlorinated Dioxins and Related Compounds. Vol. 13, Federal Environmental Agency, Austria.

75. Flint, R.W. and J. Vena (eds.) 1991. Human Health Risks From Chemical Exposure: The Great Lakes Ecosystem. Lewis Publ., Inc., Chelsea, MI. 295 p.

76. Schowengerdt, S. 1999. EPA Taken To The Task Over Pollution Data. Environmental News Network (ENN) 10/29/99 [online] URL: http://www.enn.com/news/enn-stories/1999/10/102999/chirality_6863.asp.

77. Lazaroff, C. 2000. Environmental toxins linked to Alzheimer's, Parkinson's diseases. Environment News Service (ENS), 5/8/2000. [online] URL: http://ens.lycos.com/ens/may2000/2000L-05-08-06.html.

78. Swan, S.H., E.P. Elkin, and L. Fenster. 1997. Have sperm densities declined? A reanalysis of global trend data. Environ. Health Perspec. 105(11): 1228-1232.

79. Edie Weekly Summaries. 2000. WWF says study of semen levels proves impact of endocrine disruptors. [online] URL: http://www.edie.net/news/Archive/2698.html.

80.   Environment News Service (ENS), 3/20/00. Pregnant women
       at risk from chemicals in food. [online] URL:
       http://ens.lycos.com/ens/mar2000/2000L-03-20-09.html.
81.   Environmental News Network (ENN) 5/11/99. Lake Ontario
       Fish Contaminate Breast Milk. [online] URL:
       http://www.enn.com/news/enn-stories/1999/05/051199/ont-
       fish_3129.asp.
82.   Environment News Service (ENS) 9/23/99. Millions of
       Children in the World's Largest Cities Are Exposed to Life-
       Threatening Air Pollution. [online] URL:
       http://ens.lycos.com/e-wire/Sept99/23Sept9901.html.
83.   Environment News Service, 3/1/00. High traffic streets linked
       to childhood cancer. [online] URL:
       http://ens.lycos.com/ens/mar2000/2000L-03-01-03.html.
84.   Environment News Service (ENS) 6/15/99. Pesticide Found
       in Amniotic Fluid Blocks Male Hormone. [online] URL:
       http://ens.lycos.com/ens/jun99/1999L-06-15-09.html.
85.   Environment News Service. 2000. Chemical exposures linked
       to developmental disabilities. ENS Ameriscan, 5/12/00.
       [online] URL: http://ens.lycos.com/ens/may2000/2000L-05-
       12-09.html.
86.   Kirby, A. 2000. Pollution damages intelligence. British
       Broadcast Channel News. [online] URL:
       http://news.bbc.co.uk/hi/english/sci/tech/newsid_722000/72290
       7.stm.
87.   Kaiser, J. 2000. Just how bad is dioxin? Science 228: 1941-44.
88.   Flint, R.W. and J. Vena (eds.) 1991. Human Health Risks
       From Chemical Exposure: The Great Lakes Ecosystem. Lewis
       Publ., Inc., Chelsea, MI. 295 p.
89.   Environment News Service (ENS) 2/19/99. PCBs in Texas
       Dolphins a Warning to Humans. [online] URL:
       http://ens.lycos.com/ens/fed99/1999L-02-19-03.html.
90.   Sutherland, D. 1999. Tug-o-War: Cancer Kids vs Water
       Pollution. Environment News Service (ENS) 9/27/99.

[online] URL: http://ens.lycos.com/ens/sep99/1999L-09-27-01.html.

91.   Lescroart, J. 1999. Nothing But the Truth. Random House, Inc., New York, NY. pg. 109.

92.   Williams-Hill, D., C.P. Spears, S. Prakash, G.A. Olah, T. Shamma, T. Moin, L.Y. Kim, and C.K. Hill. 1999. Mutagenicity studies of methyl-tert-butylether using the Ames tester strain TA102. Mutat. Res. 446(1): 15-21.

93.   Environment News Service (ENS), 3/9/2000. U.S. EPA Orders Oil Companies to Pay to Replace Santa Monica Water Supplies Contaminated by MTBE. [online] URL: http://ens.lycos.com/ens/mar2000/2000L-03-09-01.html.

94.   Press Release, 3/20/00. MTBE Threatens Thousands of Public Drinking Wells. American Chemical Society, Washington, DC. (Beverly Hasell - 202-872-4065).

95.   Giuliano, J.A. 1999. What Is The True Cost of Living? Healing Our World: Weekly Comment. Environment News Service (ENS) 7/19/99 [online] URL: http://ens.lycos.com/ens/jul99/1999L-07-19g.html.

96.   Robert, Karl-Henrik. 1991. Educating a Nation: The Natural Step. In Context #28, In Context Institute, Spring 1991. [online] URL: http://www.context.org/ICLIB/IC28/Robert.htm

97.   Estes, J.A., M.T. Tinker, T.M. Williams, and D.F. Doak. 1998. Killer whale predation on sea otters linking oceanic and nearshore ecosystems. Science 282: 473-476.

98.   Jacobs, J. 2000. The Nature of Economies. The Modern Library, New York, NY. pg. 145.

99.   Jacobs, J. 2000. The Nature of Economies. The Modern Library, New York, NY. pg. 46.

100.   Wackernagel. M. and W. Rees. 1996. Our Ecological Footprint. New Society Publ., Gabriola Island, BC, Canada. pg. 35.

101.   Brown, S.K. 1999. Burnt by the sun downunder. Science 285: 1647-48.

102.   Daly, H.E. 1996. Beyond Growth. Beacon Press, Boston, MA. pg. 90.

103.   McDonough, W. and M Braingart. 1998. The next industrial revolution. The Atlantic Monthly, October 1998.

104.   Jacobs, J. 2000. The Nature of Economies. The Modern Library, New York, NY. pg. 83.

105.   Maser, C. 1996. Resolving Environmental Conflict: Towards Sustainable Community Development. St. Lucie Press, Delray Beach, FL. pg. 85.

106.   Jacobs, J. 2000. The Nature of Economies. The Modern Library, New York, NY. pg. 85.

107.   Bonner, J.T. 1988. The Evolution of Complexity by Means of Natural Selection. Princeton University Press, Princeton, NJ.

108.   Environment News Service (ENS), 9/15/99. Global Eco-crisis Imminent. [online] URL: http://ens.lycos.com/ens/sep99/1999L-09-15-03.html.

109.   Daly, H.E. and J.B. Cobb. 1994. For The Common Good. Beacon Press, Boston, MA.

110.   Montague, P. 1998. Sustainable Development, Part 4. Rachel's Environment and Health Weekly #627. [online] URL: http://www.monitor.net/rachel.

111.   Black, P.E. 2000. Biosphere management: Some tools of the trade. Science 287: 235.

112.   Roach, J. 2000. Global Warming Warnings are More than Hot Air. Environmental News Network (ENN), 1/30/00. [online] URL: http://www.enn.com/news/2000/01/013/2000/climate_9521.asp.

113.   Montague, P. 1998. Sustainable Development, Part 3. Rachel's Environment and Health Weekly #626. [online] URL: http://www.monitor.net/rachel.

114.   Jacobs, J. 2000. The Nature of Economies. The Modern Library, New York, NY. pg. 130.

115.   Daly, H.E. 1996. Beyond Growth. Beacon Press, Boston, MA. pg. 93.

116. Daly, H.E. 1996. Beyond Growth. Beacon Press, Boston, MA. pg. 163.

117. Montague, P. 1998. Sustainable Development, Part 6. Rachel's Environment and Health Weekly #629. [online] URL: http://www.monitor.net/rachel.

118. Daly, H.E. 1996. Beyond Growth. Beacon Press, Boston, MA. pg. 59.

119. Monterey Bay Aquarium. 2000. Seafood Watch Chart. [online] URL: http://www.mbayaq.org/efc/efc-oc/seafood_chart.html.

120. Roach, J. 2000. What is Climate Change? Environmental News Network (ENN) Special Reports, 2/18/00. [online] URL: http://www.enn.com/specialreports/climate/what.asp.

121. Roach, J. 2000. Borehole temperatures confirm global warming. Environmental News Network (ENN), 2/17/00. [online] URL: http://www.enn.com/news/enn-stories/2000/02/02172000/borehole_10118.asp.

122. Kerr, R.A. 2000. Draft report affirms human influence. Science 288: 589-90. Mann, M.E., R.S. Bradley, and M.K. Hughes. 1999. Northern Hemisphere temperatures during the past Millenium: Inferences, uncertainties, and limitations. Geophysical Research Letters, 26: 759-762.

123. Anonymous. 1999. Global Climate Change - Overview. The Mercury's Rising 1(1): pg. 7.

124. Kerr, R.A. 2000. Draft report affirms human influence. Science 288: 589-90.

125. Anonymous. 2000. U.S. Climate Change. Washington Post Article, June 11, 2000. [Online] URL: http://www.washingtonpost.com/wp-dyn/articles/A39384-2000June11.html.

126. Heil, M. 2000. Scientists describe global climate change crisis. Environment News Service (ENS), 2/24/00. [online] URL: http://ens.lycos.com/ens/feb2000/2000L-03-24-06.html.

127. Environment News Service (ENS), 9/16/99. Global Warming May Be Triggering Super Storms. [online] URL: http://ens.lycos.com/ens/sep99/1999L-09 16-09.html.

128. Heil, M. 2000. Scientists describe global climate change crisis. Environment News Service (ENS), 2/24/00. [online] URL: http://ens.lycos.com/ens/feb2000/2000L-03-24-06.html.

129. Environment News Service (ENS), 6/4/99. Global Warming Could Raise Sea Levels in New York City. [online] URL: http://ens.lycos.com/ens/jun99/1999L-06-04-09.html.

130. Environment News Service (ENS), 6/4/99. Suffering as Planet Warms. [online] URL: http://ens.lycos.com/ens/jun99/1999L-06-04-09.html.

131. Wuethrich, B. 2000. How climate change alters rhythms in the wild. Science 287: 793.

132. Epstein, P. 1999. Climate and health. Science 285: 347-348.

133. Heil, M. 2000. Scientists describe global climate change crisis. Environment News Service (ENS), 2/24/00. [online] URL: http://ens.lycos.com/ens/feb2000/2000L-03-24-06.html.

134. Hempel, M. 2000. Tendency does not have to be destiny. Population Press 6(1): 3.

135. Maser, C. 1997. Sustainable Community Development: Principles and Concepts. St. Lucie Press, Delray Beach, FL. pg. 28.

136. Maser, C. 1997. Sustainable Community Development: Principles and Concepts. St. Lucie Press, Delray Beach, FL. pg. 196.

137. Jones, C.G., R.S. Ostfeld, M.P. Richard, E.M. Schauber, and J.O. Wolff. 1998. Chain reactions liking acorns to Gypsy Moth outbreaks and Lyme Disease. Science 279: 1023-25.

138. Maser, C. 1997. Sustainable Community Development: Principles and Concepts. St. Lucie Press, Delray Beach, FL. pg. 195.

139.   Bernard, T. and J. Young. 1997. The Ecology of Hope: Communities Collaborating for Sustainability. New Society Publ., Gabriola Island, BC, Canada. pg. 21.

140.   Maser, C. 1996. Resolving Environmental Conflict: Towards Sustainable Community Development. St. Lucie Press, Delray Beach, FL. pg. 84.

141.   Maser, C. 1997. Sustainable Community Development: Principles and Concepts. St. Lucie Press, Delray Beach, FL. pg. 54.

142.   Giuliano, J.A. 2000. Its Worse Than You Think. Healing Our World: Weekly Comment, Environment News Service (ENS), 2/14/00. [online] URL: http://ens.lycos.com/ens/feb2000/2000L-02-14g.html.

143.   Environment New Service (ENS), 7/20/99. Growing Population Faces Shrinking Water Supply. [online] URL: http://ens.lycos.com/ens/jul99/1999L-07-20-01.html.

144.   Brown, L. 1999. China's Water Crisis Linked to Global Security. Population Press 5(5): 5.

145.   Mustikhan, A. 1999. Pakistan Provinces Feud Over Water. Environment News Service (ENS), 7/28/99. [online] URL: http://ens.lycos.com/ens/jul99/1999L-07-28-02.html.

146.   Edie Weekly Summaries, 1/21/2000. Israel: Water Not Land is Key to Shepherdstown Talks. [online] URL: http://www.edie.net/news/Archive/2223.html.

147.   Montague, P. 1998. Sustainable Development, Part 2. Rachel's Environment and Health Weekly #625. [online] URL: http://www.monitor.net/rachel.

148.   Environment News Service (ENS), 10/1/98. 40% of World Deaths Environmental. [online] URL: http://ens-news.com/ens/OCT98/1998-10-01-03.html.

149.   Korten, D.C. 1995. When Corporations Rule The World. Kumarian Press, West Hartford, CT. pg. 52.

150.   Burka, K. 1999. Morocco Sounds The Alarm On Environmental
        Degradation. African News, 5/17/99. [online] URL:
        http://www.africanews.org/pana/environment/19990517/feat7.html.
151.   Mustikhan, A. 1999. Pakistan Pays Heavy Price for Eco-
        Degradation. Environment News Service (ENS), 3/10/99.
        [online] URL: http:ens.lycos.com/ens/mar99/1999L-03-10-
        01.html.
152.   Farley, M. 1999. The Grittiest Air On Earth. Population
        Press 5(5): 9-10.
153.   Lazaroff, C. 1999. Benefits of Clean Air Regs Top Costs Four
        to One. Environment News Service (ENS), 11/17/99.
        [online] URL: http://ens.lycos.com/ens/nov99/1999L-11-17-
        07.html.
154.   Montague, P. 1998. Sustainable Development, Part 5:
        Emissions Trading. Rachel's Environment and Health Weekly
        #628. [online] URL: http://www.monitor.net/rachel.
155.   Hardneer, G. 1996. Shrinking Fields: Cropland Loss in a
        World of Eight Billion. Worldwatch Institute, Washington, DC.
156.   Daly, H.E. 1996. Beyond Growth. Beacon Press, Boston,
        MA. pg. 40.
157.   Bartlett, A.A. and E.P. Lytunk. 1995. Zero growth of the
        population in the United States. Population and Environment
        16(5): 415-428.
158.   Maser, C. 1997. Sustainable Community Development:
        Principles and Concepts. St. Lucie Press, Delray Beach, FL.
        pg. 26.
159.   Jacobs, J. 2000. The Nature of Economies. The Modern
        Library, New York, NY. pg. 96.
160.   Jacobs, J. 2000. The Nature of Economies. The Modern
        Library, New York, NY. pg. 99.
161.   Daly, H.E. 1996. Beyond Growth. Beacon Press, Boston,
        MA. pg. 159.
162.   Daly, H.E. 1996. Beyond Growth. Beacon Press, Boston,
        MA. pg. 222.

163. Jacobs, J. 2000. The Nature of Economies. The Modern World Library, New York, NY. pg. 107.

164. Daly, H.E. 1996. Beyond Growth. Beacon Press, Boston, MA. pg. 207.

165. Daly, H.E. 1996. Beyond Growth. Beacon Press, Boston, MA. pg. 222.

166. Jacobs, J. 2000. The Nature of Economies. The Modern Library, New York, NY. pg. 106.

167. Shaw, R. 2000. Spare the salmon and reap the revenue, report says. Environmental News Network (ENN) 2/7/00. [online] URL: http://www.enn.com/news/enn-stories-online/02072000/salmonbenefits_9596.asp.

168. Thomas, W.L. 2000. The Age of Addiction. (in preparation for publication).

169. Thomas, W.L. 2000. The Age of Addiction. (in preparation for publication).

170. Hawken, P. 1993. The Ecology of Commerce: A Declaration of Sustainability. HarperCollins Publ., New York, NY.

171. Korten D.C. 1995. When Corporations Rule The World. Kumarian Press, West Hartford, CT. pg. 159.

172. Lyonette, Kevin, 8/2/99, personal communication.

173. Hartman, T. 1997. Last Hours of Ancient Sunlight. Mythical Books, Northfield, VT. pg. 53-54.

174. Daly, H.E. 1996. Beyond Growth. Beacon Press, Boston, MA. pg. 220.

175. Jacobs, J. 2000. The Nature of Economies. The Modern Library, New York, NY. pg. 8.

176. Jacobs, J. 2000. The Nature of Economies. The Modern Library, New York, NY. pg. 10.

177. Jacobs, J. 2000. The Nature of Economies. The Modern Library, New York, NY. pg. 31.

178. Jacobs, J. 2000. The Nature of Economies. The Modern Library, New York, NY. pg. 19.

179. Thomas, W.L. 2000. The Age of Addiction. (in preparation for publication).

180. Lescroart, J. 1999. Nothing But the Truth. Random House, Inc., New York, NY. pg. 213.

181. Robert, K. 1996. Educating a Nation: The Natural Step. In Context #28, Context Institute, pg. 10. [online] URL: http://www.context.org/ICLIB/IC28/Robert.htm.

182. Myers, D.G. 2000. Wealth, well-being, and the new American Dream. The Center for a New American Dream, Bi-monthly Column. [online] URL: http://www.newdream.org/column/2.html.

183. Myers, D.G. 2000. Wealth, well-being, and the new American Dream. The Center for a New American Dream, Bi-monthly Column. [online] URL: http://www.newdream.org/column/2.html.

184. Lynch, Damon, personal communication, 5/18/99.

185. Giuliano, J.A. 2000. It's Worse Than You Think. Environment News Service (ENS), 2/14/00, Healing Our World: Weekly Comment. [online] URL: http:ens.lycos.com/ens/feb2000/2000L-02-14g.html.

186. Cox, P.A. 2000. Will Tribal Knowledge Survive the Millennium? Science 287: 44-45.

187. Lewan, T. 1999. Television's arrival transforms tundra village. The Virginia-Pilot Newspaper, 5/23/99: A13.

188. Shoumatoff, A. 1990. The World Is Burning. Little, Brown, and Co., Boston, MA. pg. 88-89.

189. Shoumatoff, A. 1990. The World Is Burning. Little, Brown, and Co., Boston, MA. pg. 83.

190. Meadows, D.A. 1999. Building Sustainable Communities. Population Press 5(3): 12-13. Foder, E. 1999. Better Not Bigger: How To Take Control of Urban Growth and Improve Your Community. New Society Publ., Gabriola Island, BC, Canada. 175 pp.

191.   Meadows, D.A. 1999. Building Sustainable Communities. Population Press 5(3): 12-13.
192.   Giuliano, J.A. 1998. You Are Not Alone. Environment News Service (ENS), 12/30/98, Healing Our World: Weekly Comment. [online] URL: http://ens.lycos.com/ens/dec98/1998L-12-30g.html.
193.   Wackernagel. M. and W. Rees. 1996. Our Ecological Footprint. New Society Publ., Gabriola Island, BC, Canada. pg. 154.
194.   Thomas, W.L. 2000. The Age of Addiction. (in preparation for publication).
195.   Blancard, A. 1999 Darkness Peering. Bantam Books, New York, NY. pg. 174.
196.   Maser, C. 1996. Resolving Environmental Conflict: Towards Sustainable Community Development. St. Lucie Press, Delray Beach, FL. pg. 189.
197.   Maser, C. 1996. Resolving Environmental Conflict: Towards Sustainable Community Development. St. Lucie Press, Delray Beach, FL. pg. 53.
198.   Thomas, W.L. 2000. Age of Addiction. (in preparation).
199.   Maser, C. 1996. Resolving Environmental Conflict: Towards Sustainable Community Development. St. Lucie Press, Delray Beach, FL. pg. 66.
200.   Graffy, Elizabeth, USGS, Richmond, VA, 3/28/99, personal communication.
201.   Costanza, R., J. Cumberland, H. Daly, R. Goodland, and R. Norgaard. 1997. An Introduction To Ecological Economics. St. Lucie Press, Boca Raton, FL.
202.   Wackernagel. M. and W. Rees. 1996. Our Ecological Footprint. New Society Publ., Gabriola Island, BC, Canada. pg. 139.
203.   Wackernagel. M. and W. Rees. 1996. Our Ecological Footprint. New Society Publ., Gabriola Island, BC, Canada. pg. 143.

204.   Trainer, T. 1999. The Simple Plan - personal communication of ideas on simple living in Australia, 12/7/99. (e-mail: F.Trainer@unsw.edu.au).

205.   Korten, D.C. 1995. When Corporations Rule The World. Kumarian Press, West Hartford, CT. pg. 10.

206.   Holling, C.S. 1999. Lessons for sustaining ecological science and policy through the Internet. Conservation Ecology 3(2): 16 [online] URL: http://www.consecol.org/vol3/iss2/art16.

207.   Crichton, M. 1999. Ritual abuse, hot air, and missed opportunities. Science 283: 1461-1463.

208.   Anderson, C.B. 1999. Two realms and their relationships. Science 286: 907-908.

209.   Turner, M.G. and S.R. Carpenter (eds.) 1999. Commentaries: Tips and traps in interdisciplinary research. Ecosystems 2:275-307.

210.   Carpenter, S., W. Brock, and P. Hanson. 1999. Ecological and social dynamics in simple models of ecosystem management. Conservation Ecology 3(2):4. [online] URL: http://www.consecol.org/vol3/iss2/art4.

211.   Kloor, K. 2000. Returning America's forests to their natural roots. Science 287: 573-575.

212.   Holling, C.S. 1978. the spruce-budworm/forest-management problem, pg. 143-182. In: C.S. Holling (ed.), Adaptive Environmental Assessment and Management. John Wiley, New York, NY.

213.   Janseen, M.A. and S.R. Carpenter. 1999. Managing the resilience of lakes: A multi-agent modeling approach. Conservation Ecology 3(2): 15. [online] URL: http://www.consecol.org/vol3/iss2/art15.

214.   Holling, C.S. 1988. Temperate forest insect outbreaks, tropical deforestation, and migratory birds. Memoirs of the Entomological Society of Canada 146: 21-32. Peterson, G., C.R. Allen, and C.S. Holling. 1998. Ecological resilience, biodiversity, and scale. Ecosystems 1: 6-18.

215.   Allen, C.R. 2001. Ecosystems and immune systems: hierar-
        chical response provides resilience against invasions.
        Conservation Ecology 5(1): 15. [online] URL:
        http://www.consecol.org/vol5/iss1/art15.
216.   Enserink, M. 1999. Plan to quench the Everglades' thirst.
        Science 285: 180.
217.   Nicodemus Pitre Bernard, a 16 year-old on Youth Roundtable
        at the NTM meeting, Detroit, May 5, 1999, from Eunice High
        School, Eunice, LA.
218.   Maser, C. 1997. Sustainable Community Development:
        Principles and Concepts. St. Lucie Press, Delray Beach, FL.
        pg. 103.
219.   Hall, J.W. 1995. Gone Wild. Delacorte Press, New York, NY.
        pg. 80.
220.   Hempel, M. 2000. Tendency does not have to be destiny.
        Population Press 6(1): 3.
221.   Getz, W.M., L. Fortmann, D. Cumming, H. duToit, J. Hilty, R.
        Martin, M. Murphree, N. Owen-Smith, A.M. Starfield, and
        M.I. Westphal. 1999. Sustaining natural and human capital:
        villagers and scientists. Science 283: 1855-1856.
222.   Maser, C. 1997. Sustainable Community Development:
        Principles and Concepts. St. Lucie Press, Delray Beach, FL.
        pg. 66.
223.   Maser, C. 1996. Resolving Environmental Conflict: Towards
        Sustainable Community Development. St. Lucie Press, Delray
        Beach, FL. pg. 119.
224.   The Earth Charter Campaign. 1999. The Earth Charter,
        Benchmark Draft 2, April, 1999. The Earth Council,
        International Secretariat, San Jose, Costa Rica. [online] URL:
        http://www.earthcharter.org/draft.
225.   Dorin, R. 1999. Berkley School Gets Taste of Organic School
        Lunch. Environmental News Network. [online] URL:
        http://www.enn.com/food/news/9909/05/organic.lunches.

226.   Reckess, G.Z.   2000.   Gaviotas: Sustainability in an unforgiv-
       ing land.   Environmental News Network (ENN), 3/23/00.
       [online] URL: http://www.enn.com/fea-
       tures/2000/03/03232000/gaviotas_10055.asp.
227.   Reckess, G.Z.   2000.   Gaviotos: Sustainability in an unforgiv-
       ing land.   Environmental News Network (ENN), 3/23/00.
       [online] URL: http://www.enn.com/fea-
       tures/2000/03/03232000/gaviotas_10055.asp. Weisman, A.
       1998.   Gaviotas: A Village to Reinvent the World.   Chelsea
       Green Publishing Co., White River Junction, VT.   232 pp.
228.   Potts, M.   1993.   Independent Living: Living well with power
       from the sun, wind, and water.   Chelsea Green Pub. Co.,
       White River Junction, VT.   pg. 85-88.
229.   Brown, L.   2000.   U.S. farms double cropping and wind ener-
       gy.   Worldwatch Institute, New York, NY.   [online] URL:
       http://www.worldwatch.org/chairman/issue/000607.html.
230.   Kallio, N.   2000.   Powering up with renewable energy.
       Environmental News Network (ENN), 2/4/00.   [online] URL:
       http://www.enn.com/features/2000/02/02042000/windpow-
       er_9163.asp.
231.   Atkisson, A.   1999.   Believing Cassandra.   Chelsea Green
       Publishing Co., White River Junction, VT.   pg. 206.
232.   Anonymous.   1999.   Everyday Activist, Cutting CO2.
       Greenpeace Magazine, Fall, 1999.   pg. 15
233.   Myers, N.   2000.   Sustainable consumption.   Science 287:
       2419.   Environment News Service (ENS), 4/21/99.   New
       Study Finds Technological Innovation and Economic Growth
       Have Caused Continuous Improvement in U.S. Environmental
       Indicators.   [online] URL: http://ens.lycos.com/e-
       wire/April99/apr209911.html.
234.   Lazaroff, C.   1999.   Benefits of Clean Air Regs Top Costs Four
       to One.   Environment News Service (ENS), 11/17/99.
       [online] URL: http:ens.lycos.com/ens/nov99/1999L-11-17-
       07.html.

235. Ray Anderson, personal communication - National Town Meeting for a Sustainable America, Detroit MI, May 2, 1999.

236. Alexine Jackson, personal communication - National Town Meeting for a Sustainable America, Detroit MI, May 2, 1999.

237. Edwards, O. 2000. A Dangerous New Web of Communication. Sky Magazine, Delta Airlines. pg. 80-82.

238. Lahiti, T. 1998. The Agenda 21 Guide - Summary. Esam, Umea, Sweden - e-mail: torbjorn@esam.se.

239. Robert, K.H. 1991. Educating a Nation: The Natural Step. In Context #28, In Context Institute, Spring 1991: 10. [online] URL: http://www.context.org/ICLIB/IC28/Robert.htm. Gips, T. 1998. The Natural Step Four Conditions for Sustainability or "System Conditions. Alliance for Sustainability & Sustainability Associates, Minneapolis, MN. 2 pp.

240. Gips, T. 1998. The Natural Step Four Conditions for Sustainability or "System Conditions. Alliance for Sustainability & Sustainability Associates, Minneapolis, MN. 2 pp.

241. Hawken, P. 1993. The Ecology of Commerce: A Declaration of Sustainability. HarperCollins Publ., New York, NY.

242. Hartman, T. 1997. Last Hours of Ancient Sunlight. Mythical Books, Northfield, VT. pg. 53.

243. Korten, D. 1995. When Corporations Rule the World. Kumarian Press, West Hartford, CT. pg. 103

244. Sonsnowchik, K. 2000. EnvrioDesign 4 Conference. Envirospace, 3/8/00. [online] URL: http://www.envirospace.com/eco_design/view_article.asp?article_id=180.

245. Maser, C. 1997. Sustainable Community Development: Principles and Concepts. St. Lucie Press, Delray Beach, FL. pg. 101.

246. Paris, K. 1997. Environmental Indicators for Agriculture. OECD Publications #2: 22-26. Fenton P. Wilkinson,

Sustainable Options, Everson, WA; personal communication, 1/10/96. Tel. (360)966-2504.

247.   Woodruff, T.J., D.A. Axelrad, J. Caldwell, R. Morello-Frosch, and A. Rosenbaum. 1998. Public health implications of 1990 air toxics concentrations across the United States. Environ. Health Perspect. 106(5): 245-251.

248.   Environment News Service (ENS), 10/1/98. 40% of World Deaths Environmental. [online] URL: http://ens-news.com/ens/oct98/1998-10-01-02.html.

249.   Bernard, T. and J. Young. 1997. The Ecology of Hope: Communities Collaborating for Sustainability. New Society Publ., Gabriola Island, BC, Canada. pg. 43-60.

250.   Bernard, T. and J. Young. 1997. The Ecology of Hope: Communities Collaborating for Sustainability. New Society Publ., Gabriola Island, BC, Canada. pg. 43-60.

251.   Marsh, L. 1999. Environmentally sustainable lifestyles: households as partners in environmental stewardship. Personal communication from Langdon Marsh, Director, Oregon Department of Environmental Quality.

252.   Marsh, L. 1999. Environmentally sustainable lifestyles: households as partners in environmental stewardship. Personal communication from Langdon Marsh, Director, Oregon Department of Environmental Quality.

253.   Anonymous. 1999. Swedish City Builds Sustainable Urban District. Edie Weekly Summaries, 4/6/99. [online] URL: http://www.edie.net/news/Archive/1238.html.

254.   Gertler, N. 1995. Industrial Ecosystems: Developing Sustainable Industrial Structures. Massachusetts Institute of Technology, Boston, MA.

255.   Cohen-Rosenthal, E. and T.N. McGalliard. 1998. Eco-Industrial Development: The Case of the United States. Institute for Prospective Technological Studies (ITPS), The ITPS Report, Special Issue: Clean Technologies, Number 27, September 1998. [online] URL: http://www.jrc.es/iptsreport/Vol27/english/COH1E276.htm.

256.   Anonymous. 1999. Millions of children in the world's largest cities are exposed to life-threatening air pollution. Environmental News Service. [Online] URL: http://ens.lycos.com/e-wire/Sept99/23Sept9901.html.

257.   Woodruff, T.J., D.A. Axelrad, J. Caldwell, R. Morello-Frosch, and A. Rosenbaum. 1998. Public health implications of 1990 air toxics concentrations across the United States. Environ. Health Perspect. 106(5): 245-251. World Faces Cancer "Epidemic" Says Health Expert. In Asian Daily News [online] URL: http://asia.dailynews.yahoo.com/.../world_faces_cancer_epi-demic_say_health_experts.htm.

258.   Woodruff, T.J., D.A. Axelrad, J. Caldwell, R. Morello-Frosch, and A. Rosenbaum. 1998. Public health implications of 1990 air toxics concentrations across the United States. Environ. Health Perspect. 106(5): 245-251. Environment News Service (ENS), 4/23/99. Most Americans Face Air Cancer Risks.

259.   World Resource Institute. 1993. The 1993 Information Please Environmental Almanac. Houghton Mifflin, NY.

260.   UNEP GEO Team. 2000. Global Environment Outlook: Overview Geo 2000. United Nations Environment Programme, Nairobi, Kenya.

261.   Anonymous. 2001. Safe Climate web site. [Online] URL: http://www.SafeClimate.net.

262.   California Energy Commission web site: [Online] URL: http://www.energy.ca.gov.

263.   Hempel, M. 2000. Tendency does not have to be destiny. Population Press 6(1): 3.

264.   California Energy Commission web site: [Online] URL: http://www.energy.ca.gov.

265.   Environmental News Network (ENN), 7/2/99. Ozone Depleting Gases are not Natural. [online] URL: http://www.enn.com/news/enn-stories/1999/070299/ozone_4144.asp.

266.   Environmental News Network (ENN), 7/17/98. Top Power
       Sources Called Top Polluters. [online] URL:
       http://www.enn.com/enn-news
       archive/1998/07/071798/power.asp.

267.   Chiras, D. 1993. Can the human race survive the human
       race? Sustainable Futures, Spring 1993: 2.

268.   Runyan, C. 1999. The World Watch Report. Considering a
       Smaller World. Environmental News Network (ENN), 2/3/99.
       [online] URL: http://www.enn.com/enn-features-
       Archives/1999/020399/population.asp.

269.   Environment News Service (ENS), 2/4/2000. Backyard
       Burning Could be Major Source of Dioxins. [online] URL:
       http://ens.lycos.com/ens/jan2000/2000L-01-04-06.html.

270.   Woodruff, T.J., D.A. Axelrad, J. Caldwell, R. Morcello-Frosch,
       and A. Rosenbaum. 1998. Public health implication of 1990
       air toxics concentrations across the U.S. Envir. Health
       Perspectives 106: 245-251.

271.   Chiras, D. 1993. Can the human race survive the human
       race? Sustainable Futures, Spring 1993: 2.

272.   Environmental News Network (ENN), 2/6/98. Heavy Rains
       Drive Home Pollution Problem. [online] URL:
       http://www.enn.com/archives/1998/02/020698/yotostry.asp.

273.   Environmental News Network (ENN), 3/29/99. Beware,
       Don't Eat the Fish. [online] URL: http://www.enn.com/enn-
       news-archive/1999/03/032999/croaker_2349.asp.

274.   Wilson, A. and P. Yost. 2001. Buildings and the environment:
       the numbers. Environmental Building News, May 2001: 1.

275.   Vorosmarty, C.J., P. Green, J. Salisbury, and R.B. Lammers.
       2000. Global water resources: vulnerability from climate
       change and population growth. Science 289: 284-287.

276.   Anonymous. 1999. Growing population faces shrinking water
       supply. Environment News Service. [Online] URL:
       http://ens.lycos.com/ens/jul99/1999L-07-20-01.html.

277.   Environment News Service (ENS), 7/20/99. Growing
       Population Faces Shrinking Water Supply. [online] URL:
       http://ens.lycos.com/ens/jul99/1999L-07-20-01.html. Edie
       Weekly Summaries, 1/21/2000. Israel: Water, not land, is the
       key to Shepherdstown talks. [online] URL:
       http://www.edie.net/news/Archive/2223.html.
278.   Brown, L.R. 1999. China's water crisis linked to global secu-
       rity. Population Press, September/October 1999: 5.
279.   Anonymous. 2000. [Online] URL:
       http://www.newdream.org/monthly/aug00.html.
280.   UNEP GEO Team. 2000. Global Environment Outlook:
       Overview Geo 2000. United Nations Environment
       Programme, Nairobi, Kenya.
281.   Anonymous. 2000. [Online] URL:
       http://www.newdream.org/monthly/aug00.html.
282.   Collier, C. 1999. Save Some for Tomorrow. Environment
       News Service. [Online] URL: http://ens.lycos.com/e-
       wire/July99/20july9903.html.
283.   Lazaroff, C. 2000. Growing population faces diminishing
       resources. Environment News Service. [Online] URL:
       http://ens.lycos.com/ens/jan2000/2000L-01-18-06.html.
284.   Collier, C. 1999. Save Some for Tomorrow. Environment
       News Service. [Online] URL: http://ens.lycos.com/e-
       wire/July99/20july9903.html.
285.   Austin, T. 2000. Water riots seen spreading as world wells run
       dry. ENN [Online] URL: http://enn.com/news/wire-sto-
       ries/2000/09/08142000/reu_water_30512.asp.
286.   Environment News Service (ENS), 7/20/99. Growing
       Population Faces Shrinking Water Supply. [online] URL:
       http://ens.lycos.com/ens/jul99/1999L-07-20-01.html.
287.   Anonymous. 2000. Israel: Water, not land, is the key to
       Shepherdstown talks. Edie Weekly Summaries. [Online] URL:
       http://www.edie.net/news/Archive/2223.html.

288. Montague, P. 1998. Landfills are Dangerous. Rachel's
     Environment and Health Weekly, #617. pg. 2. [online] URL:
     http://www.rachel.org.

289. Sutherland, D. 1999. Tug-o-War: Cancer kids vs water pollu-
     tion. Environment News Service. [Online] URL:
     http://ens.lycos.com/ens/sep99/1999L-09-27-01.html.

290. Grossman, R. 1998. Can Corporations Be Held Accountable?
     Part I. Rachel's Environment and Health Weekly #609.
     [online] URL: http://www.rachel.org.

291. Barker, R. 1997. And the Waters Turned to Blood. Simon
     and Schuster, New York, NY. 346 pp.

292. Borenstein, S. 1998. Coastal Waters have big problems,
     Harvard study says. Buffalo Evening News, 8/25/98. (Knight
     Ridder syndication article)

293. Borenstein, S. 1998. Coastal Waters have big problems,
     Harvard study says. The Buffalo News, 8/25/98. (Knight
     Ridder syndication article)

294. Barker, R. 1997. And the Waters Turned to Blood. Simon
     and Schuster, New York, NY. 346 pp.

295. World Resource Institute. 1993. The 1993 Information Please
     Environmental Almanac. Houghton Mifflin, NY.

296. Hartman, T. 1997. Last Hours of Ancient Sunlight. Mythical
     Books, Northfield, VT. pg. 52.

297. Hartman, T. 1997. Last Hours of Ancient Sunlight. Mythical
     Books, Northfield, VT. pg. 13.

298. Wilson, A. and P. Yost. 2001. Buildings and the environment:
     the numbers. Environmental Building News, May 2001: 1.

299. Natural Resource Conservation Service. 1992. 1992 Natural
     Resources Inventory. USDA, Washington, DC
     (http://www.crcs.usda.gov)

300. Wilson, A. and P. Yost. 2001. Buildings and the environment:
     the numbers. Environmental Building News, May 2001: 1.

301. Chiras, D. 1993. Can the human race survive the human
     race? Sustainable Futures, Spring 1993: 2.

302.   Anonymous. 1999. 99% of plans to destroy US wetlands are successful. Edie Weekly Summaries. [Online] URL: http://www.edie.net/news/Archive/1426.html.

303.   Lazaroff, C. 2000. Growing population faces diminishing resources. Environment News Service. [Online] URL: http://ens.lycos.com/ens/jan2000/2000L-01-18-06.html.

304.   Steingraber, S. 1998. Living Down Stream. Vintage Press, New York, NY. pg. 7.

305.   Robbins, J. 1987. Diet For A New America. H.J. Kramer, Inc., Tiburon, CA. pg. 251.

306.   Swan, J.A. and R. Swan. 1994. Bound to the Earth. Avon Books, London, England (UK). pg. 43.

307.   Foderaro, L.W. 2001. Where apples don't pay, developers will. New York Times, June 23, 2001.

308.   Wilson, A. and P. Yost. 2001. Buildings and the environment: the numbers. Environmental Building News, May 2001: 1.

309.   Wilson, A. and P. Yost. 2001. Buildings and the environment: the numbers. Environmental Building News, May 2001: 1.

310.   Anonymous. 1999. 99% of plans to destroy US wetlands are successful. Edie Weekly Summaries. [Online] URL: http://www.edie.net/news/Archive/1426.html.

311.   Chiras, D. 1993. Can the human race survive the human race? Sustainable Futures, Spring 1993: 2.

312.   UNEP GEO Team. 2000. Global Environment Outlook: Overview Geo 2000. United Nations Environment Programme, Nairobi, Kenya.

313.   Chiras, D. 1993. Can the human race survive the human race? Sustainable Futures, Spring 1993: 2.

314.   Environmental News Network (ENN), 3/10/99. Controversial Factory Farm Controls. [online] URL: http://www.enn.com/enn-news-Archives/1999/03/031099/cleanwater_2072.asp.

315. Montague, P. 1990. Wolves Masquerading as Shepards. Rachel's Environment and Health Weekly #211, 12/11/90. pg. 2. [online] URL: http://www.rachel.org.

316. Wilson, A. and P. Yost. 2001. Buildings and the environment: the numbers. Environmental Building News, May 2001: 1.

317. Austin, T. 2000. Water riots seen spreading as world wells run dry. ENN [Online] URL: http://enn.com/news/wire-stories/2000/09/08142000/reu_water_30512.asp.

318. Schlosser, E. 1998. From the Slaughterhouse to Styrofoam, the Dirty Secrets of Fast Food. Rolling Stone 800 (11/26/98). pg. 74.

319. Schlosser, E. 1998. From the Slaughterhouse to Styrofoam, the Dirty Secrets of Fast Food. Rolling Stone 800 (11/26/98). pg. 134.

320. Colby, M. (ed.) 1998. Irradiation at a Glance, Activist Primer. Food and Water Journal, Fall 1998, Walden, VT. pg. 1.

321. Anonymous. 1999. The Growth of Antibiotic Resistant Bacteria. Food and Water Journal, Spring 1999, Walden, VT. pg. 10.

322. Environmental News Network (ENN), 9/28/98. Is Your Breakfast Genetically Engineered? [online] URL: http://www.enn.com/enn-news-archive/1998/09/092898/biotech.asp.

323. Robbins, J. 1987. Diet For A New America. H.J. Kramer, Inc., Tiburon, CA. pg. 338.

324. Anonymous. 2000. Pregnant women at risk from chemicals in food. Environment News Service (ENS). [Online] URL: http://ens.lycos.com/ens/mar2000/2000L-03-20-09.html.

325. Robbins, J. 1987. Diet For A New America. H.J. Kramer, Inc., Tiburon, CA. pg. 353.

326. Pimentel, D. 1999. How Many Americans can the Earth Support? Population Press, March/April 1999: 1-2.

327. Hartman, T. 1997. Last Hours of Ancient Sunlight. Mythical Books, Northfield, VT. pg. xxiii.

328.   Colby, M. (ed.) 1998. Irradiation at a Glance, Activist Primer. Food and Water Journal, Fall 1998, Walden, VT. pg. 1.

329.   Schlosser, E. 1998. From the Slaughterhouse to Styrofoam, the Dirty Secrets of Fast Food. Rolling Stone 800 (11/26/98). pg. 136.

330.   Hartman, T. 1997. Last Hours of Ancient Sunlight. Mythical Books, Northfield, VT. pg. 55.

331.   Colby, M. (ed.) 1998. Industry Profile: Meat Monopolies, Special Report. Food and Water Journal, Spring 1998, Walden, VT. pg. 5.

332.   Colby, M. (ed.) 1998. Industry Profile: Meat Monopolies, Special Report. Food and Water Journal, Spring 1998, Walden, VT. pg. 5.

333.   Anonymous. 1999. Lake Ontario fish contaminate breast milk. Environmental News Network (ENN). [Online] URL: http://www.enn.com/news/enn-stories/1999/05/05111999/ont-fish_3129.asp.

334.   Brown, L.R., M. Renner, and B. Halweil. 1999. Vital Signs in 1999: The Environmental Trends that are Shaping Our Future. WorldWatch Institute, Washington, DC. pg. 34. Robbins, J. 1987. Diet For A New America. H.J. Kramer, Inc., Tiburon, CA. pg. 352.

335.   Hawken, P. 1999. Monsanto and the Natural Step Revisited. Rachel's Environment & Health Weekly #676, 11/11/99. [online] URL: http://www.rachel.org.

336.   Sanjour, W. 1998. What's wrong with the EPA? Rachel's Environment and Health Weekly, #612. pg. 1. [online] URL: http://www.rachel.org. Environment News Service (ENS), 10/1/98. 40% of World Deaths Environmental. [online] URL: http://ens-news.com/ens/oct98/1998-10-01-02.html.

337.   Nader, R. 1993. The Case Against Free Trade. Island Press, Washington, DC. pg. 7.

338.   Montague, P. 1997. Toxins and Violent Crime. Rachel's Environment and Health Weekly #551. [online] URL: http://www.rachel.org.

339. Hartman, T. 1997. Last Hours of Ancient Sunlight. Mythical Books, Northfield, VT. pg. 121.
340. UNEP GEO Team. 2000. Global Environment Outlook: Overview Geo 2000. United Nations Environment Programme, Nairobi, Kenya.
341. Lazaroff, C. 2000. Polluters sully U.S. waters despite federal regulation. Environment News Service (ENS). [Online] URL: http://ens.lycos.com/ens/feb2000/2000I-02-17-07.html.
342. Gilbert, C.W. 1998. What is MCS? Blazing Tattles, May 1998. Reprinted by National Coalition for the Chemically Injured, Washington, DC.
343. Montague, P. 1997. Toxins and Violent Crime. Rachel's Environment and Health Weekly #551. [online] URL: http://www.rachel.org.
344. Hartman, T. 1997. Last Hours of Ancient Sunlight. Mythical Books, Northfield, VT. pg. 53.
345. Environmental News Network (ENN), 1/28/99. Dead Zone Burden Placed on Farmers. [online] URL: http://www.enn.com/enn-news-archive/1999/01/012899/dead-zone.asp.
346. Anonymous. 2000. Studies show pesticides damage Chesapeake Bay. Environment News Service (ENS). [Online] URL: http://ens.lycos.com/ens/aug2000/2000L-08-28-09.html.
347. Anonymous. 1998. Danger, Wear at your own Risk. Food and Water, Spring 1998. pg. 37.
348. Montague, P. 1998. One Fundamental Problem. Rachel's Environment and Health Weekly #582, 1/22/98. pg. 4. [online] URL: http://www.rachel.org.
349. Power, M. 2001. Sins of the farmers. Builder, May 2001: 272-278.
350. Korten, D. 1999. The Post Corporate World. Kumarian Press, West Hartford, CT. pg. 81
351. Rasmussen, V. 1998. Rethinking the Corporation. Food and Water Journal, Fall 1998. pg. 19.

352.   FairTax Organization. [online] URL: http://www.fairtax.org.

353.   Kolko, G. 1963. The Triumph of Conservatism. The Free Press, New York, NY.

354.   Steingraber, S. 1998. Living Down Stream. Vintage Press, New York, NY. pg. 7.

355.   Rasmussen, V. 1998. Rethinking the Corporation. Food and Water Journal, Fall 1998. pg. 19.

356.   Rasmussen, V. 1998. Rethinking the Corporation. Food and Water Journal, Fall 1998. pg. 19.

357.   Environment News Service (ENS), 10/1/98. 40% of World Deaths Environmental. [online] URL: http://ens-news.com/ens/oct98/1998-10-01-02.html.

358.   Montague, P. 1998. One Fundamental Problem. Rachel's Environment and Health Weekly #582, 1/22/98. pg. 2. [online] URL: http://www.rachel.org.

359.   Schor, J. 1998. The Overspent American: Upscaling, Downshifting, and the New Consumer. Basic Books, NY.

360.   Schor, J. 1998. The Overspent American: Upscaling, Downshifting, and the New Consumer. Basic Books, NY.

361.   Montague, P. 1995. New Strategy Focus on Corporations. Rachel's Environment and Health Weekly #309. pg. 2. [online] URL: http://www.rachel.org.

362.   African Proverb from David Creighton, personal communication, 1998.

363.   Schofield, S. 2000. I'm too busy, I can't be bothered, I'll never be able to make a difference. Envirospace. [online] URL: http:www.envirospace.com/good_practice.

364.   Curtius, M. 1999. Low-tech sewage plant is for the birds— and salmon. The Buffalo News, 1/3/99. (Syndicated from the Los Angeles Times)

365.   Bill McDonough, personal communication, National Town Meeting for a Sustainable America, Detroit, MI, May 4, 1999.

366.   McDonough, W. and M Braungart. 1998. The next industrial revolution. The Atlantic Monthly, October 1998.

367. McDonough, W. and M Braungart. 1998. The next industrial revolution. The Atlantic Monthly, October 1998.

368. Carol Browner, Administrator of the U.S. Environmental Protection Agency, personal communication, National Town Meeting for a Sustainable America, Detroit, MI, May 3, 1999.

369. Ray Anderson, personal communication, the National Town Meeting for a Sustainable America, Detroit, MI, May 2, 1999.

370. Cortese, A.D. (1992) Education for an Environmentally Sustainable Future. Environmental Science & Technology 26(6): 1009-1014.

371. Carpenter, S., W. Brock, and P. Hanson. 1999. Ecological and social dynamics in simple models of ecosystem management. Conservation Ecology 3(2): 4 [online] URL: http://www.consecol.org/vol3/iss2/art4. Kaiser, J. (ed.) 1999. Net Watch: Earth Balance Sheet. Science 286 (8 Oct. 1999): 195.

372. Janssen, M. and S.R. Carpenter. 1999. Managing the resilience of lakes: a multi-agent modeling approach. Conservation Ecology 3(2): 15 [online] URL: http://www.consecol.org/vol3/iss2/art15.

373. Fraker, T. 1999. Building Sustainable Communities. Population Press, May/June 1999, 5(4): 8-9.

374. Fraker, T. 1999. Building Sustainable Communities. Population Press, May/June 1999, 5(4): 8-9.

375. Orr, D.W. (1994) Earth in Mind: On Education, Environment, and the Human Prospect. Island Press, Washington, DC. 89 pp.

376. Jenks-Jay, N. (1995) Higher Education and the Environment: How Colleges and Universities are Responding to the Challenge of Educating Future Leaders. Nat. Assoc. Environmental Professionals News, Sept-Oct 1995: pp. 20-23.

377. Jacobson, S.K. (1995) Evaluating Impacts on Graduate Education: The Conservation and Sustainable Development Initiative. The Environmental Professional 17: 251-262.

378. Senge, P.M. 1990. The Fifth Discipline: The Art and Practic of the Learning Organization. Currency-Doubleday, New York, NY.

379.   Costanza, R. 2000. Visions of alternative (unpredictable) futures and their use in policy analysis. Conservation Ecology 4(1): 5. [online] URL: http://www.consecol.org/vol14/iss1/art5.

380.   Yankelovich, D. 1991. Coming To Public Judgement: Making Democracy Work in a Complex World. Syracuse University Press, Syracuse, NY.

381.   Costanza, R. 2000. Visions of alternative (unpredictable) futures and their use in policy analysis. Conservation Ecology 4(1): 5. [online] URL: http://www.consecol.org/vol4/iss1/art5.

382.   Costanza, R. 2000. Visions of alternative (unpredictable) futures and their use in policy analysis. Conservation Ecology 4(1): 5. [online] URL: http://www.consecol.org/vol14/iss1/art5.